POLYMER SCIENCE AND TECHNOLOGY

POLYMER SURFACES

FABRICATION, MODIFICATION AND APPLICATIONS

POLYMER SCIENCE AND TECHNOLOGY

Additional books and e-books in this series can be found on Nova's website under the Series tab.

POLYMER SCIENCE AND TECHNOLOGY

POLYMER SURFACES

FABRICATION, MODIFICATION AND APPLICATIONS

VALENTE ERASMO SILVA
EDITOR

nova
science publishers
New York

Library of Congress Cataloging-in-Publication Data

ISBN: 978-1-53614-309-6

Published by Nova Science Publishers, Inc. † New York

CONTENTS

PREFACE

Polymer surfaces having micro-nanostructures can be produced using injection molding and hot embossing, high efficiency techniques able to meet the needs of industry for the mass production of polymer parts. Micro-nanosurfaces are in great demand for multiple applications that include antipollution and self-cleaning surfaces, microlenses, dry adhesion surfaces, antireflection coatings, cell culturing and differentiation as well as superhydrophobic surfaces. *Polymer Surfaces: Fabrication, Modification and Applications* discusses the injection molding of micro-nanostructured polymer surfaces consists of three main technical steps: mold inserts, processing parameters and demolding. The authors also discuss the capabilities of various demolding methods, such as antistiction coatings, to protect and enhance the surface properties of micro-nanostructures.

The subsequent contribution focuses on biocompatible and biodegradable hollow nanoparticles prepared via the layer-by-layer deposition of polymer thin films on sacrificial templates. The review encompasses all aspects of nanocapsules from preparation through characterization and the applications as drug carriers.

A promising strategy is proposed for facilely and successively replicating randomly arranged pyramids on an etched silicon wafer to polystyrene surface. The authors propose that this fast and efficient replication strategy could be an excellent candidate for developing antireflective protective layers without complicated procedures and expensive materials.

A promising strategy was proposed for facilely and successively replicating the natural functional nanostructure of the cicada wing onto polystyrene surfaces in the concluding study. This work may direct the design of gradient wetting surfaces by mimicking the nanopillar structure of cicada wing. This strategy could aid in mimicking bio-inspired functional micro/nanostructures without complicated procedures and expensive materials.

Chapter 1 - Polymer surfaces having micro-nanostructures can be produced using injection molding and hot embossing, high efficiency techniques able to meet the needs of industry for the mass production of polymer parts. Micro-nanosurfaces are in great demand for multiple applications that include antipollution and self-cleaning surfaces, microlenses, dry adhesion surfaces, antireflection coatings, cell culturing and differentiation as well as superhydrophobic surfaces. However, the manufacturing process is not as straightforward as the scaling down of conventional injection molding. The injection molding of micro-nanostructured polymer surfaces consists of three main technical steps: mold inserts, processing parameters and demolding. Section 2 of this chapter will summarize the various techniques for fabricating mold inserts (also known as replicas, templates, inlays, stampers, etc.) and evaluate the performance of the different possible mold materials. In Section 3, the authors summarize the critical processing parameters for successful micro-injection molding and hot embossing processes. As achieving precise micro-nanostructures on polymer surfaces, requires a specific combination of processing parameters, the authors compare the effects of the various processing parameters on the quality, accuracy and precision of the final product. Parameters need to be carefully scrutinized separately for thermoplastic polymers and rubber materials. Section 4 reviews aspects related to the demolding of the prepared surface. Non-damaging demolding is the third condition for producing flawless micro-nanostructures. Micro-nanostructures may be physically damaged from adhesion and friction forces and high thermal stress. These forces that occur at the interface between the polymer surface and the mold surface are among the main constraints in the production of a high-quality textured polymer. The authors discuss the capabilities of various demolding methods, such as antistiction coatings, to protect and enhance the surface properties of micro-nanostructures. Finally, as the fabrication of superhydrophobic surfaces is one of the most in-demand micro-nanotechnology processes, Section 5 discusses the application of micro-injection molding and hot embossing techniques in manufacturing superhydrophobic surfaces.

Chapter 2 - Materials with high surface-to-volume ratio have opened new avenues in materials design for a plethora of scientific and technological areas, including energy storage and conversion, catalysis, sensing, biology and medicine. Beyond morphological aspects, the properties of high surface area materials such as polymeric nanoparticles can be tuned through surface engineering strategies, and various methods have been reported in literature so far to design polymeric nanoparticles with well-defined topology, size, chemical composition, surface properties and related biodistribution pattern and bioavailability. Beyond the widely studied core-shell nanoparticles, nanocapsules, i.e., hollow nanoparticles, have emerged as a viable and versatile platform for drug delivery applications.

This contribution will focus on biocompatible and biodegradable hollow nanoparticles prepared *via* the layer-by-layer deposition of polymer thin films on sacrificial templates. The review will encompass all aspects of nanocapsules from the preparation, through the characterization and to the applications as drug carriers. In a first section, the growth mechanism (linear vs. exponential) of LbL films on particulate templates will be discussed and rationalized based on the nature (polyelectrolyte) and characteristics (molar mass, charge state) of the polymers. The formation of nanocapsules will be presented as a function of the nature of the particulate template (metal, metal oxide, latex) and its dissolution process. Main characteristics of nanocapsules will be illustrated with respect to the experimental characterization methods. Judiciously selected examples from literature will be highlighted to demonstrate major advantages and benefits of using nanocapsules as drug delivery systems, notably in terms of drug loading and targeting ability. The last section will discuss future prospects about biodegradable and biocompatible nanocapsules in disease treatment.

Chapter 3 - A promising strategy is proposed for facilely and successively replicating randomly arranged pyramids on an etched silicon wafer to polystyrene (PS) surface. First, a nickel mold insert with negative pyramids is fabricated via combining electroless plating and subsequent electroplating with an etched silicon wafer as a template. Then, by using microinjection compression molding with the nickel mold insert as a template, the pyramids on the insert are accurately transcribed to the PS surface. The pyramids on the PS replica surface have the hemlines of about 1–7 μm and heights of about 1–4 μm. The PS replica surface exhibits a reflectance of about 5% in a wavelength range of 400–900 nm and an enhanced transmittance. That is, the surface exhibits an enhanced light trapping effect, which originates from a combination of staggered arrangement and dense distribution of the pyramids with arc-shaped top and different sizes. Especially, a thin film solar cell covered with the PS replica exhibits increased power conversion efficiency enhancement of up to 7.9% under normal incidence comparing to that covered with the PS counterpart. The solar cell covered with the PS replica also shows an excellent photovoltaic performance over a wide range of incident angles (0–70°). The proposed fast and efficient replication strategy can be an excellent candidate for developing antireflective protective layers without complicated procedures and expensive materials.

Chapter 4 - As a result of natural selection, cicadas can easily remove water droplets and dust from their wing surfaces. The fine nanostructure on the cicada wing, which exhibits hydrophobicity and antireflectivity, is carefully examined in this work. A gradient wetting state along the wing veins is revealed for the first time. That is, the Wenzel state (hydrophilic) is transitioned to the Cassie-Baxter state (superhydrophobic) from foot to apex of the wing. Superhydrophobic region exhibits a lower adhesion force to the droplets than hydrophilic region. So the accumulated droplets easily roll off from superhydrophobic to hydrophilic regions, and the light weight state on the wing can be maintained under its slight shaking.

Further, a promising strategy is proposed for facilely and successively replicating the natural functional nanostructure of the cicada wing onto polystyrene (PS) surfaces. First, a nickel replica with tapered nanopores is fabricated by combining electroless plating and subsequent electroplating with the natural cicada wing as an original template. Then, using microinjection compression molding, with the nickel replica as a template, the tapered nanopores are transcribed onto the PS surface, resulting in orderly and densely arranged nanopillars with a mean diameter of about 156 nm and a mean pitch of about 180 nm. Interestingly, the PS replica surface exhibits a water contact angle of $143° \pm 2°$ and a reflectance of about 4% in a wavelength range of 400–1000 nm. These results mean that the bionic PS replica not only inherits the nanostructure of the natural wing, but also its hydrophobic and antireflective properties. The mechanisms for the hydrophobic and antireflective properties are revealed via composite wetting interface and effective medium layer on the replica surface, respectively.

This work may direct the design of gradient wetting surfaces by mimicking the nanopillar structure on cicada wing. The proposed fast and efficient replication strategy can be an excellent candidate for mimicking bio-inspired functional micro/nanostructures without complicated procedures and expensive materials.

In: Polymer Surfaces
Editor: Valente Erasmo Silva
ISBN: 978-1-53614-309-6

Chapter 1

Fabrication of Polymer Micro-Nanostructured Surfaces: Mold Inserts, Processing Parameters and Demolding

K. Maghsoudi*, R. Jafari, G. Momen and M. Farzaneh
Department of Applied Sciences,
University of Québec in Chicoutimi (UQAC), Québec, Canada

Abstract

Polymer surfaces having micro-nanostructures can be produced using injection molding and hot embossing, high efficiency techniques able to meet the needs of industry for the mass production of polymer parts. Micro-nanosurfaces are in great demand for multiple applications that include antipollution and self-cleaning surfaces, microlenses, dry adhesion surfaces, antireflection coatings, cell culturing and differentiation as well as superhydrophobic surfaces. However, the manufacturing process is not as straightforward as the scaling down of conventional injection molding. The injection molding of micro-

* Corresponding Author Email: Khosrow.maghsoudi1@uqac.ca.

nanostructured polymer surfaces consists of three main technical steps: mold inserts, processing parameters and demolding. Section 2 of this chapter will summarize the various techniques for fabricating mold inserts (also known as replicas, templates, inlays, stampers, etc.) and evaluate the performance of the different possible mold materials. In Section 3, we summarize the critical processing parameters for successful micro-injection molding and hot embossing processes. As achieving precise micro-nanostructures on polymer surfaces, requires a specific combination of processing parameters, we compare the effects of the various processing parameters on the quality, accuracy and precision of the final product. Parameters need to be carefully scrutinized separately for thermoplastic polymers and rubber materials. Section 4 reviews aspects related to the demolding of the prepared surface. Non-damaging demolding is the third condition for producing flawless micro-nanostructures. Micro-nanostructures may be physically damaged from adhesion and friction forces and high thermal stress. These forces that occur at the interface between the polymer surface and the mold surface are among the main constraints in the production of a high-quality textured polymer. We discuss the capabilities of various demolding methods, such as antistiction coatings, to protect and enhance the surface properties of micro-nanostructures. Finally, as the fabrication of superhydrophobic surfaces is one of the most in-demand micro-nanotechnology processes, Section 5 discusses the application of micro-injection molding and hot embossing techniques in manufacturing superhydrophobic surfaces.

Keywords: micro-injection molding, hot embossing, micro-nanostructure, insert, processing parameters, demolding, durability, replication, superhydrophobicity

1. Introduction

Surfaces with micro-nanostructures and an aspect ratio (ratio of height to width) of greater than 1:1 are already used in a number of applications and interest in their use is expanding. At present, the existing applications include antireflection coatings, antipollution and self-cleaning surfaces, cell culturing and differentiation, microlenses, bioinspired non-reflective, dry adhesion surfaces and superhydrophobic surfaces [1-10]. This popularity has led to studies that aim to use the most efficient and reliable production methods for producing these surfaces at an industrial scale.

Various polymer processing techniques exist for the microfabrication of polymer materials having a specifically desired size and shape. These techniques include casting, 3D printing such as FDM and SLA, and micro-milling. However, the preferred methods in the polymer industry are injection molding and hot embossing [2, 5, 11]. This preference is related to the lower fabrication costs and faster cycle times (especially with mass production), the ability to produce multiple shapes and the simultaneous shaping of bulk and surface structures and the relatively simple ability to automate this process [8, 11-13]. As such, studies are focused on injection molding and hot embossing processes to produce microfeatured surfaces and as economically as possible [14, 15].

The injection molding of surfaces having micro-nanostructures involves the production of parts or tools having micro and/or nanostructures on their surface [16]. The required high precision in the replication of these micron or sub-micron features on the polymer surface is the key element [13, 17]. The level of precision is influenced by multiple parameters that vary in their ability to be controlled. Such parameters include feature size, shape and orientation relative to the filling flow direction, the distance of the features from the injection gate, and the aspect ratio. However, the processing conditions of the selected injection molding process, e.g., mold temperature, melt temperature, holding pressure, injection velocity, holding time and part thickness are the main influencing parameters. And of these, mold temperature, holding pressure and injection velocity are the most important in determining the quality of final replication [10]. Nonetheless, the combination of these processing parameters can produce contradictory effects that complicate their manipulation.

Micro-injection molding consists of steps similar to those of a conventional injection molding: plastification, injection (filling), packing–holding–cooling and demolding (ejection) phases (Figure 1). The filling of microfeatured cavities is much more complex than the filling of cavities in conventional injection molding [12]. The microstructures increase the cooling rate of the material in the mold due to the increase of the surface-to-volume ratio and favor a faster hardening of the molten polymer, increasing the difficulty in properly filling the micro-nanostructures before cooling of the melt [18]. Furthermore, a high-quality replication requires not only acceptable filling of the micro-nanostructures, but a non-damaging demolding of these structures as well [8]. Consequently, proper demolding is as critical of an issue as the appropriate filling.

The simple idea that molten polymers will easily fill the microstructures, cool down, solidify and be ejected without error from the mold thereby producing flawless, micro-nanofeatured surfaces on a tool, then repeated through multiple cycles, is simplistic [8]. Due to the multiple parameters influencing production, minute changes of these parameters can impact the final quality of the replicated surface.

Hot embossing (also known as direct replication) is one of the most facile, flexible and cost-effective techniques for mass-producing thermoplastic micro-nanostructured surfaces that have a high-quality replication. This is possible as thermoplastics are moldable above a certain temperature and then return to their solid state upon cooling [20]. Common thermoplastic polymers used as hot embossing substrates include polycarbonate (PC) [21, 22], polymethyl methacrylate (PMMA) [11, 22], polyvinyl chloride (PVC) [23], polytetrafluoroethylene (PTFE) [24, 25], polydimethylsiloxane (PDMS) [26] and a suite of others [21, 27]. These thermoplastic polymers are widely used in the manufacturing of membranes, optical applications (e.g., optical films), microfluidic devices, superhydrophobic surfaces, etc. [21, 28].

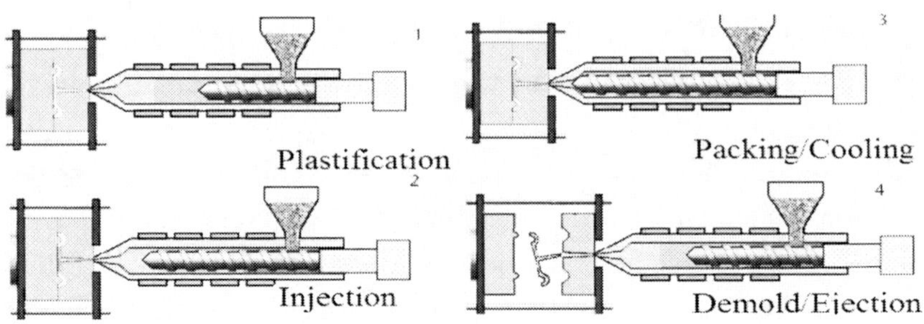

Figure 1. Schematic illustration of an injection molding system [19].

Hot embossing involves three steps: (i) creating micro-nanostructures on a template, (ii) imprinting micro-nanofeatures onto a target surface under optimal pressure and temperature conditions [23, 24] and (iii) cooling down the final product in preparation for demolding [24, 29]. As with micro-injection molding, the quality of the produced hot-embossed polymer part is highly dependent on processing conditions, i.e., processing temperature and pressure [22]. During the imprinting step, processing temperature and pressure can be increased together or individually. Evidently, the selected temperature should be above the polymer glass transition temperature T_g for the polymer material to fill all the roughness patterns on the template under the applied pressure. At temperatures higher than polymer T_g, the thermoplastic polymer enters a rubbery state where it behaves as an incompressible rubber [27]. Thermoplastic materials can undergo several heating/cooling cycles to take on the mold shape. The main advantages of hot embossing for manufacturing delicate structures are low material flow rates and the avoidance of internal stress [28].

In terms of the configuration of the molding tools, there are three different micro hot embossing systems: plate-to-plate (P2P), roll-to-roll (R2R) and roll-to-plate (R2P) [27]. P2P hot embossing is a conventional discontinuous configuration, whereas the other two are continuous making them highly efficient and suitable for large-area patterned polymeric films [27].

Although hot embossing and micro-injection molding systems are fundamentally similar, the processes do differ in a number of respects (Figure 2). The most significant difference is the temperature range to which a material is subjected [22]. The injection molding process uses polymer pellets or granules that are molten and then injected into the mold cavity as a polymer melt. Hot embossing uses a semi-finished plastic product, such as polymer plate or film [22]. Thus, in an injection molding system, polymer materials flow into the mold cavities to thereby create the micro-nanostructures. In a hot embossing system, polymer materials are softened and enter the micro-nanostructures due to the applied pressure. Consequently, the temperature and pressure ranges in the hot embossing process are considerably lower than those of the micro-injection molding system [22].

Parts having microfeatured surfaces or micro components can be made from a series of polymers. Polymers vary in terms of their rheological, thermal, mechanical, electrical and optical properties; polymers with the desired flow properties and a low viscosity at high temperatures are ideal materials for micro-injection molding [31]. Multiple polymer materials have been studied [18, 32-37] including thermoplastic polymers such as polystyrene (PS) [17], cyclic olefin copolymer (COC) [8, 31, 38, 39], poly methyl methacrylate (PMMA) [18, 38, 40], polypropylene (PP) [18, 22], polysulfone (PSU), polyoxymethylene (POM) [41], polyethylene (PE) [18], polyamide (PA) [42, 43], polyether ether ketone (PEEK) [44, 45], polycarbonate (PC) [46, 47], liquid crystal polymers (LCP) [41], polybutylene terephthalate (PBT) [48], acrylonitrile butadiene styrene (ABS) [47], polyphenyl ether (PPE) [49], and liquid silicone rubber (LSR) as a cross-linking polymer material [50]. However, polymer materials having a low elongation at the break (e.g., PMMA and COC Topas) are not appropriate for the fabrication of surfaces with a high-aspect ratio using the micro-injection molding process [51].

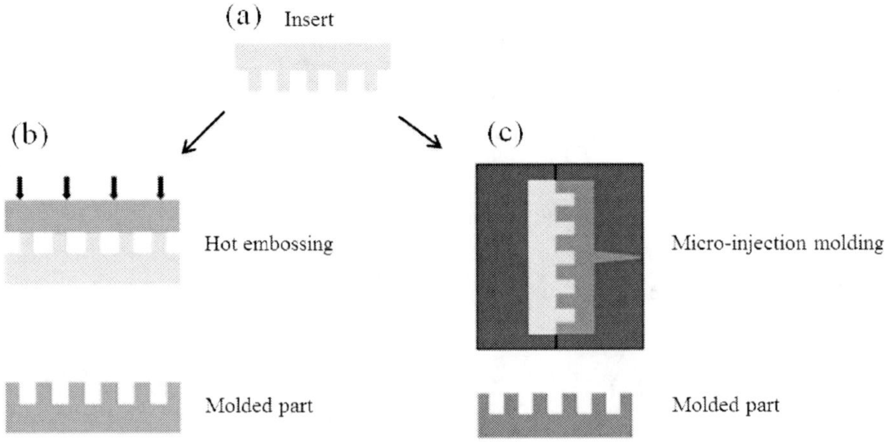

Figure 2. Schematic illustration comparing the application of (a) a microstructured insert used in the (b) hot embossing process and the (c) micro-injection molding process [30].

In this chapter, we discuss the three major steps for the fabrication of micro-nanostructured surfaces. The first step involves the selection and use of inserts to produce micro-nanofeatured surfaces within an existing mold. We assess the different materials used as inserts and discuss their respective advantages and disadvantages. The second step involves controlling the processing parameters given their critical roles in both micro-injection molding and hot embossing. We review the effects of each processing parameter on replication quality and highlight where these parameters produce contradictions in the resulting outcomes. Finally, we summarize the challenges in regard to demolding. We emphasize the role of antistiction mold coatings to ensure the non-damaging demolding of high-aspect ratio structures. An intrinsically hydrophobic polymer surface can become superhydrophobic by accepting on its surface. Given the interest in the production of superhydrophobic surfaces, in the final section of the chapter we review the use of micro-injection molding and hot embossing processes in the fabrication of superhydrophobic micro-nanostructured polymer surfaces.

2. Mold Inserts

Tool making is expensive and time-consuming. As such, use of templates is best for mass production. However, in situations requiring a limited number of examples of a form, such as when producing prototypes, there is a need for lower costs and a shorter production time. An insert (also known as a stamper or inlay) can be placed into a mold. Adding inserts to a mold is an effective approach for manufacturing structured surfaces. The intended pattern or texture is created on the template surface using the insert overlay. The pattern from the insert is then replicated onto the material. Press force is applied so that the molten material completely fills voids. Therefore, once the mold template is removed, the inverse of its pattern remains on the parent material surface [52]. This allows a faster and less expensive production of different surface features using the same mold [53]. Other advantages of the use of inserts include testing various microstructure/processing conditions without replacing the complete mold, reducing costs when replacing a worn insert rather than the entire mold, extending the life of the main mold [54] and creating patterns on the mold walls that can easily be replicated [55].

2.1. Insert Material

Replication quality is improved when the quality of the mold insert, often linked to the material used as the mold, is improved. Several materials have been used as mold inserts and have both positive and negative aspects depending on the circumstances. Mold inserts are often composed of silicon as silicon-based inserts are easy to fabricate. However, silicon can be brittle under high pressures, such as those encountered during template replication processes [5, 46, 56, 57]. Given the ease of creating patterns via mechanical machining and electroforming of creating patterns [5], metallic inserts of nickel [5, 8, 58, 59], steel [17, 41, 59], BMG (Bulk Metallic Glass) [58-60], copper [61] and aluminum [62] offer alternative materials to silicon. Metallic inserts, however, can suffer from poor replication [2] due to the higher cooling rate of the molten polymer related to the high thermal conductivity and diffusivity of metals. This higher cooling rate causes a solidified skin layer to form between the polymer melt and the mold surface, thereby preventing the molten polymer from penetrating into the micro-nanostructures.

This limitation can be avoided either by a variotherm technique (see Section 3.1 for an expanded discussion) or use other thermally insulating (low thermal conductivity) polymers as insert material (heat retardation technique). Polymers commonly used for this heat retardation technique include polyethylene terephthalate (PET), polyurethane acrylate (PUA) [12], polyimide (PI) [53], polyvinyl alcohol (PVA) [63, 64], PC [53], and

PEEK [45]. As polymeric inserts (during both the packing and filling phases) reduce the rate of cooling, the molten polymer is able to easily fill the micro-nanostructures prior to solidifying. Polymeric inserts allow melts to solidify at a slower rate (Figure 3), keeping the viscosity low for a longer period, thereby increasing the fluidity of the polymer melt [12]. As cooling can be uneven between mold portions with and without an insert [65], the inserts should have as similar thermal properties as possible to the mold to limit differences in cooling rates.

Ultraviolet (UV) cured PUA is another potential insert material. UV-cured PUA does not melt and known as a Rigiflex or “rigid-yet-soft” insert. When cured, the PUA forms strong cross-linked chemical bonds that produce a material of high rigidity (>40 MPa) and low surface energy. As Rigiflex offers a combination of both rigid and soft properties, it can be used instead of a hard or soft mold in many situations [12, 66]. UV-curable PUA resin can delay solidification near the mold surface and, as another positive aspect, facilitate the demolding of the template [12].

PVA [63] and PDMS [26], being relatively inexpensive and able to easily produce microfeatures, have been used as material for inserts in hot embossing processes. However, as these polymers suffer from problems related to proper filling (such as with metal inserts) [53], these polymers require adaptations for use in micro-injection molding processes. PI films coated with aluminum (Al) and PC have also been tested for use as insert materials. PI films produced an outstanding replication of microstructures even after undergoing more than 1000 replication cycles. However, PC inserts would deform quickly as they could not endure the heat transferred from the polymer melt [53].

In addition to metals and polymers, inserts composed of quartz [53] and gallium arsenide (GaAs) have been attempted [46]. Using a steel mold, Yoon et al. [46] placed silicon and GaAs inserts, then added a 0.4-mm-thick PTFE sheet between the two inserts to prevent damage to the inserts due to the high pressures involved.

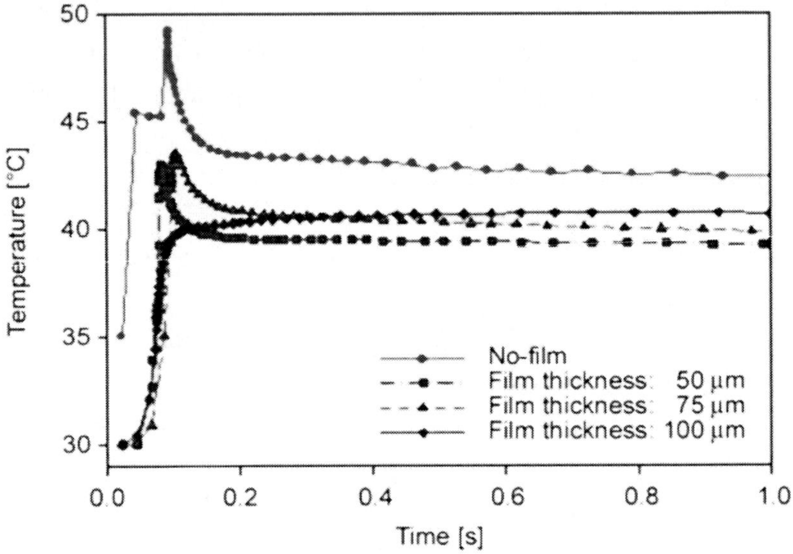

Figure 3. Changes in mold surface temperature over time, with and without a polymeric insert [12].

Hybrid inserts are composed of a layer of polymer and a layer of metal. The polymer layer is an epoxy, polyimide or polyether-ether-ketone. The metal layer ensures there is no chemical reaction between the melt and the polymer layer. The hybrid approach has been suggested for use as inserts [53, 67] as the cooling rate of the melt is lower and the temperature of the melt remains above T_g for a longer period due to the lower heat conductivity of the polymer layer. These conditions improve the filling of the mold and limit the formation of a solid skin layer while cooling. Furthermore, hybrid inserts produce a stretching effect resulting in nanopillars that are 40% higher than expected from the original insert [53]. This stretching effect helps produce the desired micro-nanostructured surfaces.

PDMS is widely used for hot embossing micro-nanostructured templates [68-72]. Via a laser ablation technique, designed patterns can be created on the PDMS surface as a template for hot embossing of amorphous polyethylene terephthalate (APET), for example [68]. The obtained replication fidelity using a PDMS template was remarkably higher than that of a steel mold template. Furthermore, elastomeric templates benefit from the uniform contact with the material surface and facile demolding (due to low surface energy). To improve PDMS stiffness and hardness as a hot embossing template, the amount of curing agent can be manipulated and there can be thermal aging of the PDMS upon curing. A combination of these two adjustments increased the stiffness (150%) and hardness (32%) of the PDMS template [73].

A rapid heating-cooling cycle can be achieved using a micro-patterned silicon stamper coated and ~45 nm thick carbide-bonded graphene network [74]. This graphene coating, due to its high thermal and electrical conductivity and low surface resistivity, can produce heating rates of 5–10°C/s allowing a polymer surface to reach its T_g within only 10 s. Therefore, this method could lead to a better replication quality, shorter cycle times and very low energy consumption [74].

2.2. Insert Fabrication Methods

Mold inserts are fabricated through a variety of techniques including (I) LIGA based (lithography, electroplating, molding) technologies (LIGA, UV-LIGA, IB-LIGA, EB-LIGA) [29, 75-78]; (II) 3D micro-machining regrouping micro electrical discharge machining (μEDM), micro mechanical milling and electrochemical machining (ECM) using ultra-short pulses [14, 59, 79]; (III) silicon wet etching [80]; (IV) deep reaction ion etching (DRIE) [2]; (V) thick deep UV resists; (VI) laser ablation [67, 79, 81-84]; and (VII) plasma treatment [2, 85-88], micro-drilling and micro-turning [89].

The μEDM approach relies on high temperature metals or alloys for fabricating the mold inserts to produce metals having a strength >200 MPa [79] The anode-cathode system is first submerged into a dielectric fluid and a high-voltage current is run between the cathode tool and anodic electrode to remove metal. For the die-sinking EDM process, a fluidic microfeature is milled into the surface of a graphite/copper electrode.

A stainless steel workpiece is then eroded using a thermal process via a controlled electric spark [59].

Of the numerous techniques for creating micro-nanotextures on metal surfaces, such as chemical etching, electro-chemical treatments and grit-blasting, it is laser ablation or laser milling that has attracted much interest due to its lower contamination and greater reproducibility [90]. A reliable approach for producing inserts out of engineering materials [25, 81, 91], laser milling occurs via material at a specific location being evaporated by a high-intensity and concentrated laser beam (Figure 4). Specific patterns are etched by moving either the laser beam or the substrate in x-y directions (see [92] and [93] for more details).

Depending on the achievable precision, the laser technique is divided to nanosecond, picosecond and femtosecond laser. Several factors determine the final micro-nanostructures, such as laser fluence, the number of laser shots and the gaseous environment [90]. Laser-induced surface texturing has been studied for a number of different materials, e.g., titanium, stainless steel, aluminum, copper, silicon, etc. [90, 94, 95]. However, the micro- and nanosurface features of these metals differed considerably when formed using a femtosecond laser. These differences can be ascribed to variable electron-phonon coupling strength as well as to thermal conductivity [94, 96]. Different metal surfaces therefore alter the replicated surfaces.

Figure 4. A laser-milled structure showing surface roughness and the resultant debris [82].

When laser milling is coupled with 3D CAD/CAM techniques, sharkskin-like surfaces can be produced on stainless steel inserts [81]. The laser milling process is accomplished by a series of random hatchings inside each layer and a clear demarcation of edges. This leads to a favorable surface integrity and edge definition. Laser ablation, by altering the chemical and structural features on the surface [97], can generate superhydrophobicity and superoleophilicity directly on the nanocomposite PDMS thin films [97] and polyimide surfaces [98].

2.3. Types of Micro-Nanostructures

The forms of produced micro-nanostructures are wide-ranging and include micro-nanochannels with different cross-sections, e.g., cylindrical, rectangular [60, 77, 99-101], micro-nanopillars [13, 17, 44, 53, 102, 103], micro-nanohairs [104, 105], micro-pyramid [21] and square meshes [106]. Peng et al. [27] provide a detailed summary of the fabrication of microstructures—also discussing microstructure type, size and aspect ratios—through the hot embossing process.

The flow and behavior of a polymer melt is altered by the presence of microfeatures on the surface [107, 108]. The viscosity of polymer melts, for example, is much greater when microstructures are on a surface relative to a non-structured surface. Young et al. [101] used a rheological model to show that at a constant temperature, an increase in the channel radius significantly increases the filling distance and filling velocity. Furthermore, as the surface area to volume ratio decreases, the cooling rate is greatly reduced.

In regard to the specific surface area ($m^2/m^3 = m^{-1}$) of different structures, the filling capacity of microchannels (triangle and rectangular cross-sections) was most difficult relative to all other shapes [68]. Circular structures had the greatest filled length due to the low specific surface area. Of three features (square, circular and hexagonal) for a hot embossing process by PDMS of a microfeatured mold, square features had the greatest obtained height percentage [68] (Figure 5).

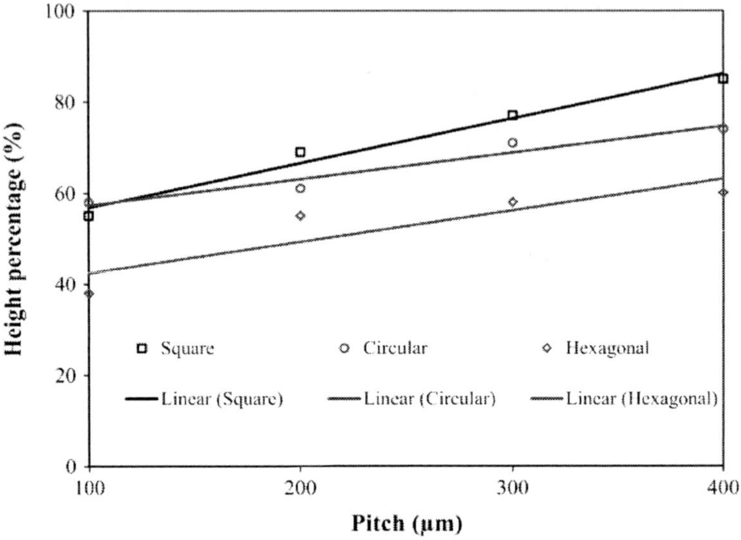

Figure 5. Height percentage of the three features embossed using a PDMS mold [68].

A more accurate filling occurs when channels are oriented parallel to the direction of main flow as opposed to channels perpendicular to flow [109, 110]. An increased aspect ratio of the microfeatures will decrease replication quality as the polymer melt solidifies faster reducing the percentage of structures being replicated [111, 112]. An optimal design of aspect ratio must therefore favor both an appropriate filling and reliable replication.

2.4. Insert Durability

The durability of the inserts over multiple molding cycles is essential to justify their use in mass production. Conventional high-strength hard inserts, such as steel, tend to be durable than polymeric inserts [12]. For nonmetal inserts, the durability of cross-linked polymers (e.g., photocurable epoxy) is better than plastic polymers (e.g., PEEK) over the course of the replication process. Cross-linked polymers increase design precision during fabrication [53]. Stormonth-Darling et al. [53] ran a single hybrid insert through more than 2000 cycles. The hybrid insert showed few signs of wear and damage. However the depth of some features was reduced. This reduced depth was primarily from polymer residues accumulating on the insert, and the original depth of the microfeatures was regained when the insert was washed with N-methyl-2-pyrrolidone (NMP).

Durability tests for BMG inserts [60] produced a number of scratches although no cracks on the surface after 10 000 molding cycles. After 20 000 molding cycles, however, the roughness of the BMG inserts increased 10-fold relative to the roughness after 10 000 cycles, and there was a marked number of cracks on the insert surface. For an anodized aluminum oxide template, micro-nanostructures on the template surface remained intact even after hundreds of molding trials [51]. The brittleness of silicon could hinder its use as an insert material. However, it is misleading to compare the fracture toughness (K_{IC}) of a silicon wafer with that of metals (0.95 $MPa{\cdot}m^{1/2}$ and 15–150 $MPa{\cdot}m^{1/2}$, respectively). Most ceramic materials have a high compressive strength rather than a high fracture toughness [46]. As such, the use of silicon remains promising, assuming it is manipulated properly. A metallic coating having a similar thermal expansion coefficient as a silicon wafer could improve the compressive durability of the silicon insert. A test of this combination showed that silicon could withstand an injection pressure of 50 MPa [46]. Furthermore, acceptable results were obtained with a PTFE sheet reinforcing a silicon wafer, even after 3000 cycles [46]. Overall, there is an improved durability and resistance to wear for inserts having a hard coating material [43, 113, 114].

Thus, the mass production of accurate replicates involves a good choice of inserts based on adequate filling, selection of durable insert material, design of an appropriate surface features and a use of the optimal fabrication technique. Following the selection of an appropriate insert, the most fundamental step for both micro-injection molding and hot embossing systems is determining the parameters of processing.

3. Processing Parameters

Micro-injection molding begins as pressure drives molten polymer into the cavity. The molten polymer initially experiences conditions of low flow resistance. However, upon arriving at the microfeatures, there is a marked increase in resistance, thereby requiring increased pressure to force the melt to fill the microfeatures [17]. Given the rapid heat loss of the melt due to contact with the cavity walls, there can be the premature formation of a solidified layer. Melt is prevented from flowing into the micro-nanostructures due to insufficient back pressure. Possible solutions to countering this situation include increased mold temperature or injection velocity or increasing both injection and holding pressure together [115].

The replication process continues until the packing phase [8, 12]. Although the pressure applied to the melt during the initial contact between the molten polymer and the mold surface is negligible, it can increase during the packing phase and push the molten polymer into the micro-nanostructures during this phase [8, 10]. However, this depends on the temperature of the polymer melt. If the melt temperature is less than the polymer T_g, increased pressure during the packing stage produces minimal effect on replication. The formation of a skin layer on the surface of the melt—where it touches the cold cavity—limits the flow of the molten polymer [8, 102].

Although processing conditions have a major effect on the hot embossing process, it is much simpler than for micro-injection molding systems. For the latter, the critical element is to achieve the polymer glass transition temperature (T_g). Above T_g, polymer deformation becomes much more facile and two main filling mechanisms occur: pressure-driven viscous flow and surface-tension-driven capillary flow [21]. The latter is attributed to the formation of polymer mounds, while the former is the simple flow of the polymer materials from the edges into the template cavities [40]. For microstructures to be faithfully replicated on the PMMA surface, temperature, the polymer viscosity, applied pressure and the time of embossing are controlled [40].

Average height and uniformity of the micro-nanostructures are also important for a high-quality replication. To achieve optimal uniformity of shape and height of microfeatures, it is essential to ensure a high holding pressure, high mold temperature and an even distribution of cavity pressure. Use of efficient rapid heat cycle molding techniques and a thicker substrate can achieve these required conditions [116]. Injection velocity is the most debated parameter. Injection velocity is the speed at which polymer is injected into the mold cavity during the injection phase. Debate involves whether higher or lower injection velocities improve replication quality [116].

An efficient filling (and more economical process design) can be achieved using conventional processing parameters at the micro-injection stage and then increasing temperature and pressure to ensure a high-filling ratio [117]. However, this technique is highly questionable in the case of rubber injection molding as the mold temperature is much lower for rubber injection than that for plastic injection molding. Thus, cross-link density and cross-linking reaction rates are increased when the temperature is increased after the filling stage. This produces undesirable conditions for micro-nanostructure filling.

As such, an optimal replication requires a combination of different processing parameters. A design of experiments (DoE) method has been used in a number of studies to identify these optimal parameter combinations [17, 31, 39, 116, 118]. A DoE method uses experimental data that is acquired in a controlled manner.

The significant and non-significant factors affecting a process are determined, and the behavior of injection molding of micro-nanostructures is carefully monitored. All of the main processing parameters are significant including mold temperature, injection velocity and holding pressure [116]. Thus, to better understanding the effect of various processing parameters on replication quality and determine the optimal processing conditions, providing a process window could be useful [18].

3.1. Mold Temperature

Increasing mold temperature is generally considered as the most practical means of improving the filling quality of microstructures having a high-aspect ratio [17, 31, 53, 76, 116, 119]. At higher temperatures, plastic polymer viscosity is reduced favoring easier flow of melt into the cavity. Optimal filling will only occur if the melt has enough fluidity to efficiently fill the microstructures. Thus, mold temperatures should be equal or greater to the softening temperature of the polymer and the temperature of the polymer melt is above its T_g [8, 17]. Most micro-nanostructure filling occurs during the packing stage. Therefore, polymer melt temperatures must remain high enough at this stage. When the mold temperature is equal to the polymer T_g, the acquired aspect ratio is 1, a value that is not ideal for forming micro-nanostructures. Higher mold temperatures, i.e., temperatures above T_g of the polymer, produce more acceptable aspect ratios [8].

The high surface-to-volume ratio in the micro-nanostructured cavity causes the temperature of the polymer melt to decreases abruptly immediately upon entering the cavity, favoring the formation of the skin layer [56]. Therefore, the mold temperature must be maintained at a high enough temperature to prevent the forming of a skin layer. Su et al. [56] recommend a mold temperature 30–40°C above the polymer T_g to ensure an efficient filling and to avoid the development of the skin layer. When the temperature is high enough (200°C), the depth-to-opening ratio (h/d) is not influenced by the radial location. High mold temperatures inhibit the fast cooling and the formation of a solid layer. Thus, the microfeatures are filled efficiently [31, 76, 102].

Increasing the mold temperature to decrease polymer viscosity has been discussed rheologically [18]. Based on two basic rheological equations (*Eq. 1 and 2*) for polymer materials, as the thickness of the cavity (h) decreases, the required shear strain rate ($\dot{\gamma}_{av}$) increases to maintain a constant average velocity (v_{av}).

$$\dot{\gamma}_{av} = 2v_{av}/h \tag{1}$$

$$\tau = \eta\dot{\gamma} \tag{2}$$

Therefore, based on the classic stress-strain equation (*Eq. 2*), which indicates the relation among shear stress (τ), polymer viscosity (η) and shear strain ($\dot{\gamma}$), shear stress in the cavity increases with an increase in shear strain and thus, a higher filling resistance for the polymer. As a result, mold temperatures greater than the glass temperature of the polymer could fulfill requirements for a successful microstructure replication. There is a direct relation between the acquired aspect ratio of the micro-nanostructures and mold temperature. As mold temperature increases, the achievable aspect ratio increases (Figure 6) [18, 110, 120, 121].

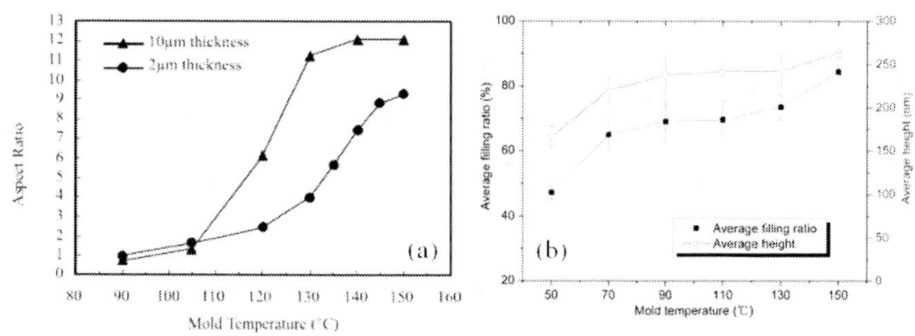

Figure 6. Effects of mold temperature on (a) the achievable aspect ratio of the micro-walls at two different thicknesses [18], and (b) the replication quality of PMMA columns vertical to the flow direction [110].

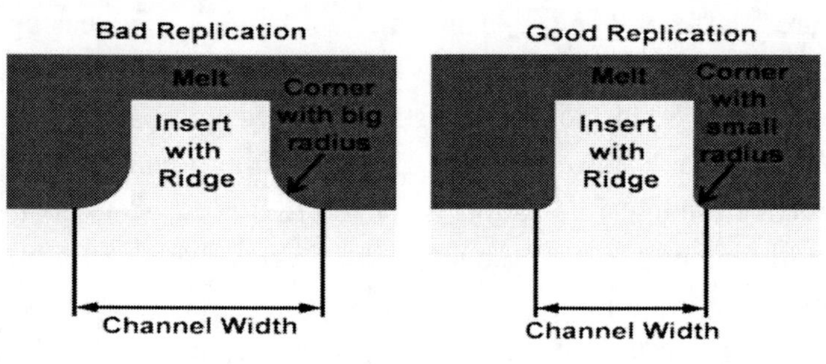

Figure 7. Schematic of a cross-section through a molded channel and the insert ridge showing rounded upper channel corners on the moldings corresponding to poor and good replication [123].

Although mold temperature and the injection velocity are viewed as the most important processing parameters for influencing polymer viscosity [102], increased mold temperature allows both injection pressure and injection velocity to be reduced [56, 122]. Thus, in an injection system, if the mold temperature is high enough, injection pressure and injection velocity would not need to be increased. A recent study investigated the failure of the polymer melt to completely fill corners, resulting in rounded upper corners in microchannels injection molding [123]. These rounded corners can be used as an index of replication quality. With poor replication, channel width increases, while good replication has inserts with a narrow channel width (Figure 7). Mold insert temperature, melt pressure and cooling time are the most influencing parameters. Both delayed and early cooling lead to reduced replication quality and, as would be expected, increased mold insert temperature leads to better replication [123, 124]. However, to achieve a complete filling, high melt pressure is also required, and this high pressure should be applied before the formation of the solid layer [123].

Many injection molding systems are used for the assembly of polymer parts. In a two-component assembly, the bonding strength is highly important. Mold temperature during the assembly process affects the morphological structure and resulting bonding strength [125]. Increasing mold temperature leads to an increase in bonding strength.

Mold and melt temperature are often deemed as the most important parameters affecting micro–nanopattern quality. However, only mold temperature affects micro–nanopattern height. Increasing mold temperature by 50% can produce a >300% increase in the average height of micropillars [17]. In the packing stage, a lower viscosity polymer melt allows an easier and more efficient filling of micro-nanostructures. This improvement in height replication is related to a reduced viscosity due to higher mold temperature and reduced elastic effects during the filling and demolding stages. With a high injection velocity, the induced shear stress leads to an elastic relaxation after removal of the manufactured part from the mold. Increasing mold temperature prevents this undesirable effect and reduces viscosity without provoking an elastic effect [55].

The use of novel types of materials and the adjustment of process parameters in micro-injection molding processes has been explored. Flexible polymeric materials rather than rigid thermoplastics are of great interest. One example is the development of thermoplastic elastomers (TPEs) used in sensor applications as micro-suspension rings [126]. Indeed, the dimensional precision of the outer (OD) and inner diameter (ID) of the manufactured micro-ring is critical. In the case of the OD, mold temperature and holding pressure had a strong influence on the response value with an increase in both parameters having a positive effect on OD [126]. Holding pressure had no considerable effect on ID. Mold and melt temperatures have significant effects on ID differently. Increased mold temperatures decrease ID while increased melt temperatures increase it. Similarly, embossing temperature is recognized as the most significant factor influencing the filling ratio [127]. Increasing embossing temperature from 51.3°C to 95.9°C almost doubled the effective filling ratio. Heating time was the least significant factor affecting the filling ratio [127].

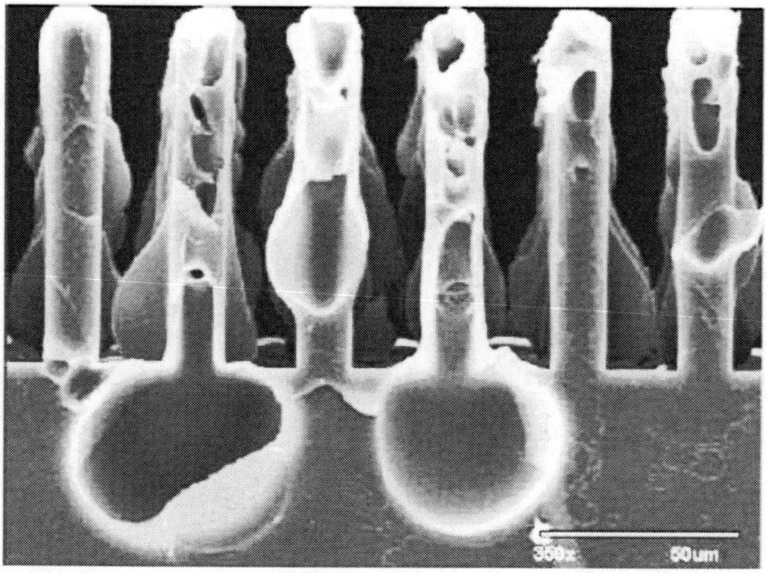

Figure 8. The effect of excessively high mold temperature on PMMA micro-walls [18].

Thermoplastic polymer recovery in the micro-hot embossing process is a less studied yet important phenomenon [21]. Polymer recovery happens after complete filling and causes a reduction in replication quality. Polymer recovery leads to a height reduction of 21%, 18% and 16%, depending on the microstructure type. Increased temperature and embossing time could offer a solution for polymer recovery. However, the consequences of applying these solutions include an increased cycle time and warping issues. Therefore, interfacial bonding between the polymer material and the cavity wall was introduced to prevent polymer recovery [21].

Nonetheless, there are some negative effects related to increased mold temperature. Higher temperatures lengthen cooling times and, thus, increases cycle times, not ideal for industrial needs [34, 76]. Furthermore, at higher mold temperatures, higher adhesive and/or frictional forces complicates demolding, reducing the replication quality of micro-nanostructures [8]. A marked increase in mold temperatures can also degrade the polymers [56, 75]. Furthermore, mold temperatures cannot be increased without limit. From SEM imagery of a PMMA micro-wall, Liou et al. [18] illustrated how the high temperatures cause gas to be produced by the polymer material, leading to damaged molding marked by multiple voids (Figure 8).

Thus, two temperature-related issues affect the replication quality of injection molded micro-nanostructures. First, during the processing stage, molten polymer fails to fill completely the high-aspect ratio micro-nanostructures. Increased mold temperature to above that of the polymer T_g and, therefore, decreasing polymer viscosity is one possible solution [8]. The second issue is the significant interfacial adhesion between the polymer and the mold substrate that is observed during demolding (see Section 4).

3.1.1. Variotherm Heating System

The formation of a premature solid layer is a major issue facing micro-nanostructure injection molding systems. As discussed above, this skin layer restrains the melt from completely filling the high-aspect ratio micro-nanostructures. And while increasing the mold temperature offers a possible solution, this also unfortunately lengthens cycle time. A preferred solution is to have a cold mold during the cooling stage and a hot mold during the injection stage [79]. This is the basis of the so-called "variotherm" system as opposed to classical approaches [10].

The variotherm process employs several methods to heat molds including various heat transfer phenomena, e.g., convection, radiation, conduction and induction [128]. In a variotherm process, the mold is heated up to the polymer T_g, and then the filling stage begins. The temperature is kept higher than polymer T_g for the polymer melt to fill up the cavity efficiently and to hinder formation of the solid layer on the neighboring cavity wall. Immediately after the packing stage is finished, the mold is rapidly cooled down leading to solidify the polymer melt. Following demolding, the mold is heated up again to be ready for the next cycle [129]. As heating and cooling rates play key roles in the variotherm process by affecting molding cycle time and injection molding efficiency, a precisely controlled mold temperature is crucial for this technology.

Rapid heat cycle molding (RHCM) is a new variotherm micro-injection molding technology that uses numerical thermal analysis based on ANSYS software. Hot steam and cold water are used as heating and cooling media capable of heating the mold up to the desired temperature in 40 s and cooling down the shaped polymer in about 34.5 s for demolding [129], much faster changes in temperature than standard variotherm technology. Moreover, the final equilibrium temperature of the mold surface was increased in the heating cycle and decreased in the cooling cycle, thereby improving the efficiency of this new variotherm mold design [129].

Overall, the advantages of the variotherm heating system include improving component quality, processing control and feature replication, increasing the polymer flow path and fluidity of the polymer, decreasing residual stresses, altering molecular orientation and flow resistance, eliminating the formation of weld lines and short shots as well as reducing material degradation [10, 79, 128, 130, 131]. Variotherm processes can compensate for an increased cycle time of a traditional heating system, while not producing an unfavorable solid film layer; however this technique, due to the frequent and rapid heating-cooling cycles, decreases mold lifetimes and also requires expensive equipment [12, 79].

3.2. Melt Temperature

Overall, the effect of melt temperature on replication quality is similar to the effects of mold temperature. As melt temperature increases, the viscosity of thermoplastic polymers decreases, improving the filling of the micro-nanostructures [76, 132]. Therefore, high melt temperatures are advantageous for achieving a high-quality micro-nanostructure. High melt temperatures, at the level of the micro-nanostructures, help the holding pressure force molten materials into these structures [8]. The increase of the melt temperature also reduces the thickness of the solid skin layers [102]. The highest possible barrel temperature, which is considered as the melt temperature, is needed to keep the materials in their flow state so that they can efficiently fill the cavity surface structures [17].

The interaction of melt and mold temperatures may be crucial for the micro-injection molding of micro-nanostructures. Higher mold temperatures coupled to melts having different temperature produce variable replication outcomes. Higher mold temperatures improve replication quality when melt temperatures are low. Quality is reduced when both mold and melt temperatures are high. Surfaces producing the highest contact angle have processing conditions of low temperature melt/high temperature mold [50].

For rubber injection molding, melt temperature had a very different impact. Cross-linking rates increase with temperature. Thus, in rubber injection, the temperature of the raw material (uncured rubber) must be kept as low as possible before cavity filling to prevent cross-linking. When the rubber material had filled the cavity, in-situ curing can then occur to produce the final cured rubber. However, if mold temperatures are high or the temperature of the raw rubber is above the temperature for which cross-linking bonds begin to form, non-ideal rubber curing happens. Under such conditions, i.e., uncontrolled curing before filling of the cavity, it becomes impossible for the rubber material to fill the micro-nanostructures. Hopmann et al. [50], studying the micro-injection molding of LSR, found that low melt temperature leads to high viscosity and, in turn, high stress. This generated pressure can be useful for forcing rubber into the structures. Furthermore, maximum heights for micro-nanostructures occurred at lower melt temperatures.

However, melt temperature can also have minimal impact on replication quality [53]. For example, Stormonth-Darling et al. found that a 10 °C increase in melt temperature did not significantly influence the achieved height of structures [55]. Increased melt temperature can also increase the degradation rate of the polymer during processing [34].

3.3. Processing Pressure

Packing (or holding) pressure is the main force that drives polymer melt into the micro-nanostructures of the cavity surface. Molten polymer fills the micro-nanostructures in the packing stage [8, 76] and, regardless of the mold height and melt temperature, great amounts of pressure are needed to push the melt into the nano-microstructures. Furthermore, this pressure can compensate for the trapped air pressure and polymer shrinkage [39, 76].

The effect of holding pressure is highly dependent on the distance of the structures from the injection gate. Moving away from the injection gate, holding pressure rapidly decreases due to frictional shear forces from flow resistance. In many respects, replication quality of the final product is related to the pressure at the points farthest from the injection gate [17]. There is a remarkable interaction between injection velocity and holding pressure. Variation in holding pressure markedly affects feature height at high injection velocities. At high pressure levels, feature height can be 50% greater [116]. However, higher injection velocities also result in a non-uniform pressure distribution that increases the standard deviation of height values.

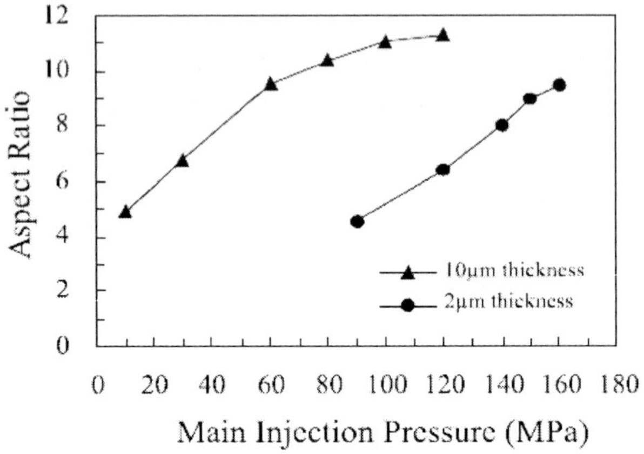

Figure 9. Effects of main injection pressure on the obtained aspect ratio of the micro-walls with two different thicknesses [18].

In general, the cavity filling time is much longer than the solidifying time of the injected material near the mold surface. Accordingly, the formed skin layer between the bulk polymer and the cavity surface counteracts the influence of packing pressure [17]. Maintaining the clamping force is useful as long as the melt is in a flow state. Once the skin layer is formed, maintaining this pressure has no effect on replication quality. Therefore, holding time is not usually considered to be a critical factor in the injection molding of micro-nanostructured surfaces [31, 116]. Where packing time is considered to be an effective factor on the average filling ratio, increasing packing time allows more molten polymer to enter the micro-nanostructures, and thereby compensating for volumetric shrinkage during the filling step [110].

Injection pressure is a term commonly used in the micro-injection molding process and refers to the pressure by which the materials are injected into the mold cavity. Evidently, injection pressure affects the flowability of polymer melt and the replication quality [133, 134]. The replicated aspect ratio increases with a higher main injection pressure, but this increasing trend is not monotonic (Figure 9). The rate of increase for the achievable aspect ratio tends to flatten as the injection pressure increases [18]. However, some researchers consider injection pressure to be a secondary factor affecting the filling capacity of the micro-nanostructures [135].

Embossing pressure and roller speed affect the replication quality of micropatterns. In micropattern fabrication by R2R micro hot embossing, better results are achieved using higher embossing pressures and higher roller speed [27, 136].

3.4. Injection Velocity

Injection velocity is one of the most important yet most debated processing parameters. Injection velocity can enhance the melt flow in the cavity thereby affecting the replication of features, especially for those located away from the injection gate [17, 132]. A higher injection velocity improves replication quality [12, 63, 75, 79, 137, 138]. First, a higher injection velocity produces higher shear rates and shear stress; thus, following polymer rheology, the viscosity of the polymer melt will be lower. Second, a higher injection velocity reduces the cooling rate during mold filling. Furthermore, as injection velocity increases, there is decreased contact time between the material and the cavity cold surface; therefore, the forming of a skin layer, before the melt fills the structures, is prevented and fewer short shots are produced.

Injection velocity is particularly important for rubber injection molding. In this approach, mold temperature is greater than melt temperature, thus the polymer material is in contact with a hot cavity rather than a cold cavity in the filling stage. A rubber compound in contact with a hot cavity surface will produce more cross-links over a shorter period. This cured skin prevents the filling of micro-nanostructures by other polymer melts and results in poor replication quality. Premature scorching should not be used to obtain an acceptable filling. A high injection velocity is the only factor that ensures a satisfactory filling in rubber injection molding as a longer filling time inevitably leads to premature scorching [50]. When injection velocity is high enough, the induced internal pressure in the cavity pushes the molten material into the microfeatures at the desired level. On the other hand, there may be dissipative heating of material due to a high injection velocity [50]. This induced heating, which results from the induced shearing by the high velocity of the polymer material, can generate a high melt temperature. For rubber injection molding, increased melt temperatures lead to undesirable consequences. As such, the selection of an injection velocity during the rubber injection process strongly influences the quality of the final product.

Song et al. [102] tested the influence of injection velocity on replication quality. They tested three injection velocities (low, intermediate and high) and assessed the replication quality of large-scale micropillar arrays of variable diameters. At low injection velocities, the mean height of the micropillars was lower. For intermediate injection velocities, due to the similarity of the melt temperature at both the beginning and the end of the filling process, the acquired aspect ratio was constant and parallel. Finally, at the highest velocities, micropillar height remained constant in the onward and the opposite directions. They concluded that injection velocity and the mold temperature should be as high as possible.

Undoubtedly, complete filling is not related entirely to an increase in injection velocity. If a high injection velocity cannot ensure a complete filling, it can at least ensure a more uniform and homogeneous molding of the structures, an aspect that has a great impact on the water contact angle of a surface [50]. According to Yokoi et al. [119], the use of a high injection velocity is the best means of avoiding the formation of solidified thin layer. They used ultra-high-speed injection molding (UHSIM) at a high injection rate. UHSIM can inject the polymer into the mold at a rate of 1000–2000 mm/s. In comparison, a conventional injection molding process has an injection rate of about 50–100 mm/s [119]. Increased injection velocities also show some contradictory results. Higher injection velocities can result in improved replication via shear thinning—a phenomenon of increased shear stress and decreased polymer viscosity. On the other hand, this generated shear stress can lead to an induced elastic stress residual in the polymeric parts after demolding. To prevent this, high mold temperatures can decrease viscosity without inducing shear thinning and a subsequent elastic stress relaxation [55, 139]. Replication depth and injection velocity present a non-monotonic behavior [13]. An increase in the injection flow rate up to 4 cm^3/s will increase replication depth; yet, above these values, replication depth decreases [13]. A minimal injection velocity is required to obtain a high-aspect ratio and desirable replication quality [116]. As injection velocity increases, there is a marked increase in the trapped air pressure between the polymer melt and the cavity surface leading to a lower surface quality in the produced material [65]. To avoid this problem, a complete air evacuation can be achieved using an active vacuum-based approach [18, 140].

Although high injection velocities improve the filling of micro-nanostructures, there can be poor surface quality and edge definition in demolded parts [132]. An enhanced feature size has been observed at low injection velocities [55]. The replication of all feature sizes (except for the smallest sizes–100 nm) was enhanced as injection velocity decreased. This may be due to the similarity between the hot embossing process and the micro-injection molding process at a zero-injection velocity. However, this idea needs to be verified. Therefore, selection of the appropriate injection velocity is highly dependent on other processing conditions, including types of polymer, mold and melt temperature and holding pressure.

3.5. Part Thickness

There remains much debate in regard to the role of thickness of the molded part (also known as substrate) as an influence on the replication of micro-nanofeature surfaces in injection molding. Some maintain that a decrease in part thickness improves filling capability, while other studies demonstrate that replication quality deteriorates as part thickness is reduced. The geometry of the substrate affects the pressure profile. This, in turn, affects the filling capability of the molten polymer. As a substrate is thicker than the microfeatures, the substrate can easily alter the pressure profile and thereby affect replication quality [17]. Thicker substrates reduce replication quality by limiting increases of in-cavity pressure during the molding process. In contrast, thinner substrates favor an increase of in-cavity pressure that tends to increase filling height.

However, the role of thickness remains complicated. As wall thickness decreases, the injection molding of microfeatures becomes more difficult. Increased substrate thickness leads to a uniform distribution of cavity pressure in the holding stage. Therefore, to improve replication quality, a greater wall thickness becomes essential [18, 116]. A thicker substrate is also recommended to increase the filling rate and prevent the creation of a solid thin layer [141]. Given a constant injection rate, as the thickness of the molded part increases, cavity filling time also increases [17]. All this together demonstrates that comprehensive part design requires a consideration of the role of thickness and its influence on replication quality.

3.6. Filling Distance

The distance of micro-nanofeatures from the injection gate also influences replication quality. In general, replication quality is better for structures located closer to the injection gate. Filling height quickly decreases moving away from the gate [116]. Processing parameters are also involved as a higher holding pressure closer to the injection gate favors a greater replication quality for micro-nanofeatures close to the gate.

Radial location can influence the replicated aspect ratio [56]. The obtained aspect ratio is not uniform between different cavity sizes. Replicated microstructures near the gate have higher depth-to-opening ratios (h/d) due to the higher local temperature and pressure. As radial location increases, the aspect ratio decreases, except in the areas close to the edge of the mold. Along the edge, there is an increase in generated back pressure favoring an increase in the depth-to-opening ratio even beyond that of structures in the central area [10].

A significant height reduction was observed in the hot embossing of micropillar arrays and micro-prism arrays on a circular PC surface [21]. This height reduction occurred as micropillars were positioned far away from the center point to the circumferential area due to the pressure reduction. However, this phenomenon was also attributed to polymer recovery. Other studies, however, have shown an opposite, positive relationship between replication quality and distance from the gate [56, 132, 142]. Although the distance of the structures from the gate is believed to be an important, albeit debatable, parameter affecting the replication quality, distance between micro-nanostructures has no significant effect on the filling efficiency [132].

3.7. Vacuum Mold Venting (Air Evacuation)

The presence of air trapped in the mold cavity creates a major problem during molding by producing voids and bubbles in the bulk or surface portion of the final assembled part, incomplete filling (known as short shot), poor appearance, surface combustion (or the diesel effect), burn marks and in some cases permanent damage to the mold [65, 119, 143].

The presence of air in the mold is inevitable. It is essential to eliminate air from the mold cavity particularly from "blind" cavities [144]. This issue is much more crucial for the injection molding of nanostructures [119]. Numerous solutions for evacuating trapped air have been proposed [18, 75, 145-147]. The resulting defects, e.g., short filling (leading to a hollow defect) and rough tip surface, produced by the loss of a vacuum are evident in Figure 10 [144].

Conventional venting, through the use of vents on the mold walls to let trapped air escape, and air evacuation, via pumps to remove air, are the two main methods for avoiding the formation of voids. In standard injection molding, conventional venting is efficient. However, this approach is not feasible for micro-injection molding parts and molds; conventional venting leads to unfavorable structural changes in the micro-molded materials [56]. Thus, to evacuate trapped air in micro-injection molds, the most efficient means is a combination of a vacuum pump and small holes in the mold [147].

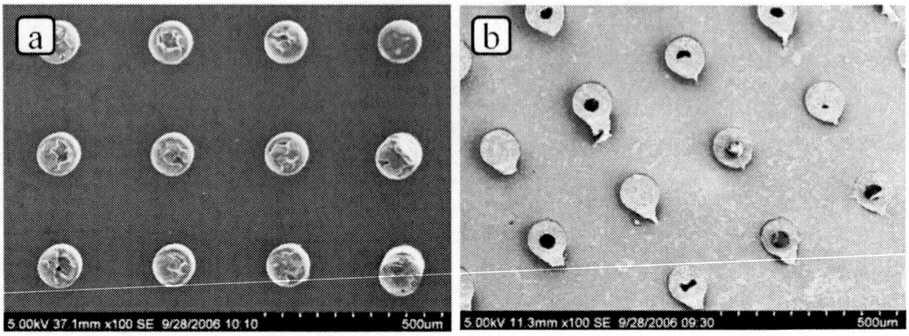

Figure 10. SEM images of PP micro columns molded without vacuum causing (a) rough tip and (b) hollow defect (PP columns array after cut) [144].

Yokoi et al. [119] used the transcription ratio (TR, defined as the ratio of V-groove depth within the molded sample to that of the insert) to compare filling quality with variable processing parameters, both with and without vacuum pumping. Vacuum pumping increased the aspect ratio. Vacuum venting for micro-injection molding of PMMA also improves the depth ratio [148] and has a greater impact when the mold temperature and injection velocity are low, and the feature geometry is narrow [148]. In contrast, a 16% decrease in average microstructure height was observed in a study of molding with air evacuation. Mold temperature was the parameter most influencing aspect ratio, as vacuum mold venting evacuated warm air within the cavity, decreasing the mold temperature and, as a consequence, reducing replication quality [65, 116]. In another experiment, air loss from the cavity resulted in a mean temperature decrease of about 7% [149]. This temperature decrease is greater when the mold temperature is higher, regardless of injection velocity [149]. These adverse effects are particularly striking for polymers that are sensitive to temperature change, such as PS [116]. Air evacuation should therefore be avoided as replication of micro-nanostructure height appears not optimal.

In addition to these main processing parameters, other less-studied parameters may impact micro-nanostructures. These other parameters include maximum ejection temperature [147], metering size, main injection time [18], runner type (hot or cold) [79] and movement of the injection plunger to control the holding pressure [150]. These parameters directly or indirectly affect the main aforementioned parameters.

4. DEMOLDING

High-quality replication is dependent on mold filling, processing conditions and a flawless demolding. Demolding must be achieved so that the micro-nanopatterns retain their mold heights, and the micro-nanostructures can resist the potentially damaging demolding forces. Poor design and demolding of the manufactured piece from the mold can result in severe damage to the micro-nanofeatures on the surface, failure of the molded features on the polymer surface and also can reduce the lifetime of the mold [8, 151, 152]. Therefore, demolding is a key factor in commercializing the manufacture of micro-nanostructured polymer surfaces [29].

Numerous factors need to be considered to ensure a proper demolding, including polymer selection and mold conditions as well as processing parameters and part design [153]. More specifically, demolding is affected by the applied forces to the polymer surface, differences in the thermal conduction coefficient between the polymer material and the mold metal, generated forces due the shrinkage of the polymer during the solidification stage, process parameters, cavity shape and material, molding material and geometry, features shape and aspect ratio, etc. [151].

There are currently two main demolding methods. The first approach uses antistiction coatings, which are chemical surface agents applied to facilitate demolding. The second is a mechanical approach applying ejectors that includes use of pins, blades, rings, sleeves and stripper blades [31, 151, 154]. The selection of the appropriate method is not always straightforward as both have some major disadvantages. The probable harmful effects on human health from chemical demolding agents restrict their use in medical and some microfluidic applications [79]. Mechanical ejectors can permanently damage a fabricated part. Proper design of the produced tool and the optimal placing of ejector pins are both necessary to avoid damage to the polymeric parts, particularly when the geometry of the piece is complicated [31, 39, 151]. Given these limitations of conventional demolding techniques, novel methods have emerged and are being tested, such as ultrasonic vibration [79] and the reduction the surface roughness to decrease the coefficient of friction of the mold surface [155].

According to Saha et al. [114], there are eight different specific aspects affecting replication of micro-nanostructure surfaces in both hot embossing and micro-injection molding processes: (i) the coefficient of friction, (ii) the surface energy of both polymer and mold surface, (iii) the roughness especially along the sidewalls, (iv) the slanting angle of the structures, (v) the chemical bonding at the interface, (vi) the thermal expansion coefficient, (vii) the environment and resulting corrosion and (viii) the replication temperature.

In general, demolding is comprised of two different forces: adhesion and friction. Adhesion is the dominant force at the intersection of the microfeature bottom and the top surface of the polymer material. Adhesion force is a combination of chemical adhesion, electrostatic adhesion and capillary attraction [156]. Friction is the force produced by the movement of the molded polymer inside the feature along the walls of the microfeature. The force acts in the opposite direction to the ejection movement [151].

Processing parameters can have a major effect on demolding forces. The increase in both holding pressure and mold temperature leads to a decrease in the demolding force [156-158]. Increased holding pressure can reduce the shrinkage and consequently decrease demolding forces. Longer cooling times lead to higher demolding forces [151], while melt temperature and injection pressure also influence demolding forces [151, 159]. Based on a Taguchi DoE method, mold temperature is the dominant parameter determining demolding forces [151, 160], while melt temperature has a moderate effect and packing pressure and time have negligible effects [151]. In terms of the physical effects, as part thickness increases, the demolding force decreases but not monotonically [151].

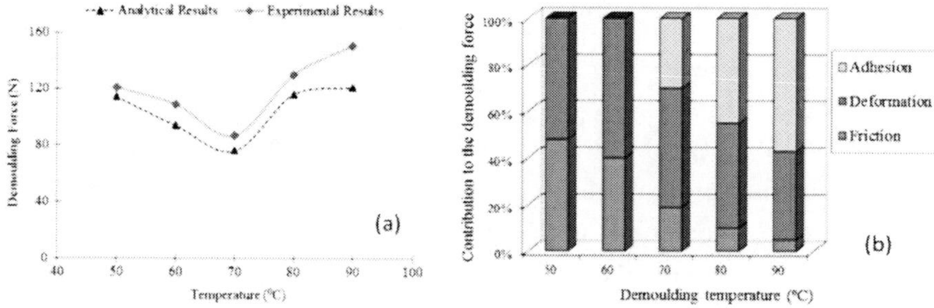

Figure 11. (a) Demolding force of hot embossing as a function of demolding temperature, and (b) contribution of various demolding forces types versus demolding temperature [161].

Demolding temperatures could affect demolding forces for the hot embossing of PMMA [161] (Figure 11(a)). Testing an aluminum mold and various demolding temperatures, Omar et al. [161] found an optimal temperature where the demolding force was minimal; increasing or decreasing temperatures from this optimal temperature increased the demolding force. The contribution of different acting forces, including adhesion, deformation and friction during demolding varied with temperature (Figure 11(b)). At higher temperatures, the adhesion force dominates, while at lower temperatures it has no contribution. The deformation force plays the main role in demolding forces at lowest temperatures [161].

Demolding temperature is also related to cooling time, as shorter cooling times occur with a higher demolding temperature. Increased cooling time leads to higher filling ratio, and a short cooling time (i.e., higher demolding temperature) lowers the polymer tensile strength [110]. This may lead to fractured micro-nanostructures during demolding. Consequently, to obtain a high average filling ratio and flawless demolding, a longer cooling time is favored.

Ultrasound vibration of the mold core can decrease interface friction during the ejection step. During the demolding of PS, COC and POM, ultrasound vibration reduced the mean value of the demolding force for all mentioned polymers [162]. This reduction was even greater (16%) in the case of the micro-injection molding of PS. As an external force, this vibration leads to a localized heat built-up that can soften the polymer, due to the reduction in storage modulus, and thereby reducing its mechanical interlocking [162].

Micro-milling (μm) and micro-electro discharge machining (μEDM) are two machining techniques for creating variable mold textures. In a series of tests, where both machining technologies led to nearly similar roughness (R_a) factors, the ejection force for each technique is highly dependent on the mold temperature [156]. At higher mold temperature (a standard condition for micro-injection molding), the ejection force of the μEDM is greater due to a better filling of the μEDM surface topography by the molten polymer at higher temperatures. Furthermore, R_a was found to be too general considering the elaboration of ejection friction. As such, a complete description of surface topography and roughness is required to be provided by a profiler machine.

4.1. Antistiction Coating

Mold coatings, such as titanium nitride (TiN) and ceramics, were conventionally used to limit erosion, wear, thermal fatigue, etc., which all shorten the life of a mold [114]. Increased aspect ratio, structural complexity and the need for high precision led to a greater demand for antistiction and low friction coatings [114]. Although release agents are usually judged as being suitable for larger components, they can be applied to micro-replication processes [114, 152]. The quality of the molded surface and the uniformity of the nanostructures can be improved through use of antistiction coating, while preventing the surfaces from having microcracks, polymer residues, stretching and tearing during part ejection [8, 163]. Self-assembled monolayers (SAM) are often used as antistiction coatings for aggressive conditions that require low surface energy coatings having a superior tribological performance [13, 114]. The most common coatings are fluorocarbon- or hydrocarbon-based coating materials such as fluorocarbonsilanes [164-167], fluorocarbon phosphoric acid [8], fluorocarbons [29] or alkanethiols [8]. Perfluorodecyltrichlorosilane

(FDTS) has also been used as an effective antistiction coating for both hot embossing and micro-injection molding.

The appropriate mold coating for hot embossing and micro-injection molding should possess a series of properties: (I) a low friction coefficient, (II) low adhesion, (III) high wear resistance, (IV) chemical inertness, (V) thermal stability, (VI) an ability to withstand high pressure, (VII) non-toxicity, (VIII) an ability to adhere well to base material and (IX) an ability to maintain all of the above functionalities even for a layer a few atomic thick [114].

Equation 3 [8] calculates the maximum tensile stress σ_{pillar} on a cylindrical pillar. The friction force F_{fric} is calculated based on the shear stress σ_{macro} required for detaching polymeric parts from a mold surface. The cross-sectional area A_{pillar} of a cylindrical pillar can be calculated by its diameter w and height h:

$$\sigma_{pillar} = F_{fric}/A_{pillar} = 4\sigma_{macro}h/w \tag{3}$$

As an example, from *Eq. 3*, the demolding forces for a nanopillar having a diameter of 40 nm and a height of 110 nm would have a maximum tensile stress of demolding of 110 MPa for a FDTS-coated mold. For an unmodified mold, the demolding stress would be greater than 550 MPa [8]. Moreover, the average height of the nanopillars for samples produced with FDTS-coated inserts is greater than those produced without the antistiction coating, regardless of mold temperature [8]. For successful demolding, tensile forces must overcome the aggregate polymer/mold adhesive and friction forces. If demolding stress exceeds the ultimate tensile strength of bulk polymers, the molded polymeric structures may fracture [8].

Fluorosilanes are common antistiction coatings, although some modification of their use may be required [2, 168] as no coating is suitable for all molding processes. Techniques, such as water contact angle measurement can provide a means of evaluating and choosing the ideal antistiction coating. Stormonth-Darling et al. [2] review possible surface coatings that may be used to improve replication quality.

The consistency of the antistiction coating is a critical factor to have successful demolding. Inconsistent coating produces a stretching effect produced from the friction between the mold surface at the bottom of the structure and the polymer material. When the antistiction layer is consistent, the stretching effect is absent and ideal demolding of the material is possible [2]. The stretching effect during demolding has also been observed when the polymer melt neighboring the mold surface remained above T_g [53]. Stormonth-Darling et al. [2] proposed a "success rate" to identify the abundance of perfectly formed pillars. The success rate is defined as a ratio between the total numbers of perfectly formed features to the total number of visible features on a given surface. Thus, broken or low height features (generated by improper filling) are not considered as successfully replicated pillars. Successful replication occurs when the features do not overlap, and the feature has an adequate contrast to its neighbors. In a test of different coatings, three fluorinated coatings, TPFS-only, SiO_2+TPFS, and Si_3N_4+TPFS, demonstrated a greater stretching capability and a defined success rate generally >80%. However, metal surfaces, such as Ni and Ti, could not successfully fill ultra-high-aspect ratio structures, although their success rate in terms of stretching capability was close to 100% [2].

Other tests with antistiction coatings include applying a 10-nm-thick plasma-polymerized fluorocarbon-based coating on a nickel mold surface. This coating significantly improved replication quality as it did not reduce the replication depth and did not undergo any degradation even after hundreds of injection molding cycles [13]. Micro-injection molding of thermoplastic polyurethane (TPU) with silicon tooling was coupled with the use of TPFS as the antistiction coating material, applied using a vapor self-assembled monolayer (VSAM) approach [139]. With this combination, the antistiction coating enhanced the effect of mold temperature as higher mold temperatures increased the depth ratio by more than two-fold, while the depth ratio was less altered at lower mold temperatures. Although the application of antistiction coatings tends to improve demolding and replication of micro-nanostructures [55], some negative results have been produced. Antistiction coatings have actually limited the filling of nanostructures in situations where there was insufficient wetting of the coating by the polymer melt [13]. For high-quality replication, demolding without damage or altering of micro-nanostructures is as important a step as the complete filling of the mold during processing [53].

5. Superhydrophobic Micro-Nanostructured Polymer Surfaces

For a material to possess superhydrophobic properties, it should first demonstrate hydrophobic characteristics that are due to the non-polar groups on the material's surface. Moreover, its surface must consist of micro-nanofeatures that roughen the surface [169]. There are two main regimes for describing the wetting behavior influenced by surface roughness and chemical composition. The first is the Wenzel state where the water droplet penetrates into the surface features. The second is called the Cassie-Baxter state where the water droplet rests on top of the surface micro-nanofeatures and air pockets are trapped underneath the water droplet [169]. Among the most commonly used techniques to fabricate micro-nanostructured polymer surfaces are micro-injection molding and hot embossing processes. These two techniques are the simplest and most efficient means for producing superhydrophobic surfaces.

In this section, we study the application of micro-injection molding and hot embossing techniques to the making of superhydrophobic surfaces. This section focuses solely on template replication base systems for the fabrication of superhydrophobic micro-nanostructured polymer surfaces. This section does not discuss methods where a template replication base route is not used, e.g., spray coating [3, 170-172], spin coating [173], plasma surface treatments [174, 175], coating of SAM [51, 176-178], dip coating [179, 180] or specific case study techniques [97, 181, 182].

5.1. Wetting Theories

The wetting property of a surface is defined by Young's equation, which describes the wetting of a solid surface by contact angle value (θ):

$$\cos\theta = \frac{\gamma_{SA} - \gamma_{SL}}{\gamma_{LA}} \tag{4}$$

where γ_{SA}, γ_{SL} and γ_{LA} represent, respectively, solid-air, solid-liquid and liquid-air interfacial tensions. These surface tension forces correspond to the static contact angle of a liquid droplet on a solid surface [183]. However, the basis of wetting characterization of rough surfaces relies mainly on Wenzel [184] and Cassie-Baxter [185] theories. Wenzel theory describes the homogenous wetting regime as follows:

$$\cos\theta_W = r\cos\theta \tag{5}$$

where θ_W is the Wenzel contact angle and r is the roughness factor, which is defined as the ratio of the actual surface area to the projected surface area. From this equation, when the surface is hydrophobic ($\theta > 90^o$), increasing the roughness leads to an increased contact angle [186].

For a heterogeneous wetting regime (a surface composed of two fractions), the contact angle is given by the Cassie-Baxter equation:

$$\cos\theta_{CB} = f\cos\theta + f - 1 \tag{6}$$

where θ_{CB} is the Cassie-Baxter contact angle, f is the fractional area defined as the ratio of the solid area wetted by water to the total solid area [186, 187]. By incorporation of the roughness factor of the wet area (r_f), a modified Cassie-Baxter equation is obtained [186]:

$$\cos\theta_{CB} = r_f f\cos\theta + f - 1 \tag{7}$$

It is worth noting that when $f = 1$ and $r_f = r$, the Cassie-Baxter equation becomes the Wenzel equation [186].

5.2. Superhydrophobic Polymer Surfaces by Hot Embossing

Superhydrophobic PC possessing a water contact angle of 155° has been fabricated using silicon wafer templates [169]. As the wafer temperature increased (ranging from 150°C to 175°C), the water contact angle also increased. Although the T_g of this PC was around 147°C, contact angles increase considerably at temperatures higher than 155°C. This is attributed to the difference between silicon wafer surface temperature and the actual temperature of the PC surface due mainly to the heat dissipation at the interface. However, a surface having water contact angles of greater than 150° (this can be considered as a superhydrophobic surface) occurred at temperatures above 170°C. This pattern results from the stiffness of the PC at various temperatures. As discussed in Section 3.1, the viscosity of the molten polymer decreases with temperature. Therefore, at temperatures lower than 170°C, the PC is not soft enough to penetrate easily into the microtextures and create high micropillars. This causes the surface to act within a Wenzel state. However, at the temperatures above 170°C, the PC is soft enough to enter the microtextures under the applied weight. Therefore, the micropillars are high enough (> 6 μm) for the surface to act within the Cassie-Baxter regime [169].

Using femtosecond laser ablation, Toosi et al. [23] imprinted topographical stainless steel (SS) micro-nanostructured patterns onto the surface of a range of thermoplastic polymers, which included high-density polyethylene (HDPE), polylactic acid (PLA), and medical PVC. Temperatures were slightly above the polymer melting points to speed up the embossing process and to minimize the deformation due to elastic recovery and relaxation. The imprinting pressures were in the range of 3–12 MPa. The PLA and HDPE produced water contact angles (CA) above 160° and a contact angle hysteresis (CAH) of less than 10°. However, the presence of hydrophilic plasticizers in the medical PVC generated a "petal effect." The petal effect represents the wettability behavior of a surface having both a high CA and high CAH. Depending on the settings of the laser ablating parameters (such as fluence, scanning speed, and beam overlap), three different morphologies were obtained on the surface of the SS: a paraboloidal pattern, a triple-roughness structure and a cauliflower pattern.

The wetting properties of HDPE surfaces were measured using different concentrations of various organic solutions, such as methanol, acetone and 2-propanol. The micro-fibrous HDPE surfaces produced a CA as high as 170° and a CAH as low as 3°. The superoleophobic behavior of the imprinted HDPE surfaces was confirmed using low surface tension liquids.

In addition to hot embossing, ultraviolet (UV) embossing can also be used to fabricate microfeatured polymer surfaces [188, 189]. This technique makes it possible to fabricate hydrophobic thermosetting resin surfaces. A UV-curable polyurethane acrylate (PUA) was cured using UV exposure on an electroplated nickel template. The same template was used for the hot embossing of polypropylene (PP). Comparing hot and UV embossing with the produced template, CA was maximized when the diameters of the cylindrical micropillars were optimal. The static water CA of the flat PP and the flat UV-cured PUA resin were 98° and 78°, respectively. The CA of the hot-embossed PP surface was higher than that of the UV-embossed PUA, confirming Wenzel's theory that creating roughness on a hydrophobic surface makes this surface more hydrophobic. However, as the flat UV-cured PUA was hydrophilic, the obtained hydrophobicity could not be explained using Wenzel's theory but rather by Cassie-Baxter's equation.

Using a simple and inexpensive fabrication process, superhydrophobic polymer films have been produced that demonstrate superior abrasion resistance and water pressure stability [190]. The process consists of polyethylene film lamination over a woven wire mesh template. After peeling off the mesh, a three-dimensional structure is created on the polymer surface to produce superhydrophobic properties for the surface (CA of 160º and sliding angle of 5º). Altering the parameters of lamination temperature, pressure and time influenced on the resultant superhydrophobicity. At temperatures lower than 120ºC, even under high pressure and a longer time, the static contact angle was 125º as the temperature was insufficient to produce a polymer film soft enough to penetrate into the mesh structures. However, at a temperature of 120ºC, which is greater than the temperature of polymer crystalline melt, the resulting CA was 160º, under identical pressure and time as the lower temperature surface [190].

Superhydrophobic surfaces can be manufactured via the hot embossing of PE with a microscale steel gauze [191]. Post-treatment by a cold radiofrequency plasma of tetrafluoromethane achieved both a superhydrophobic and superoleophobic (superomniphobic) surface. Such unique wetting properties were obtained due to the creation of hierarchical structures following the plasma treatment.

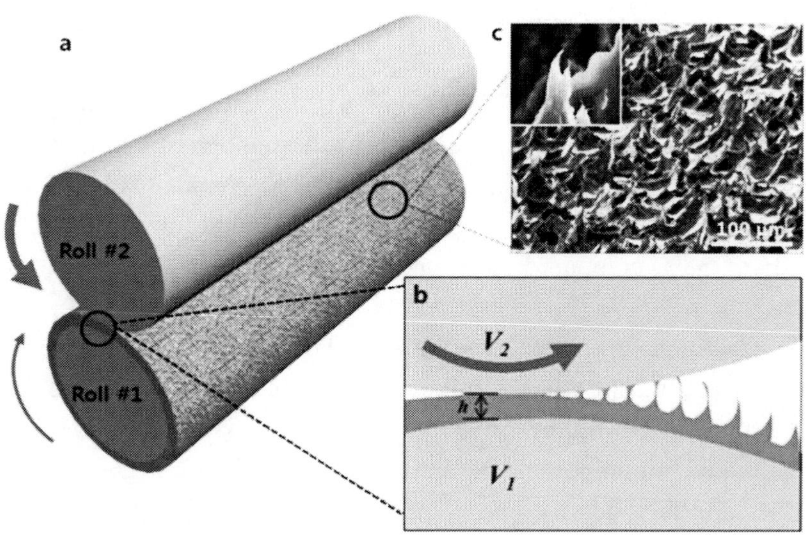

Figure 12. (a) Schematic illustration of the double-roll machine. (b) Inverted rotating rolls produced capillary bridges leading to the creation of patterns. (c) A SEM image of the patterned surface. (Inset: individual MWCNTs at the end of the pattern tip) [192].

Hot embossing has some drawbacks including a large number of cycles to produce high-volume products, a number of processing steps, a limit to the fabricable area, etc. As such, there are attempts to industrialize the techniques used for the manufacturing of micro-nanostructured surfaces [192]. A double roll–based process was employed to fabricate the sharkskin-like patterned multiwalled carbon nanotubes (MWCNTs) and PDMS composites. According to Park et al. [192], this method does not require any patterned template or stamp. The key factor for achieving a randomly structured pattern over the entire film is the ratio of roll velocities (V_2/V_1) (Figure 12). Optimization of the surface patterns is based on the effective shear rate $\eta = \Delta V/h$ between the two rolls, where $\Delta V = V_1 - V_2$ and h is the thickness of the film. An optimal shear rate of -82 s^{-1} (meaning $V_2 > V_1$) produced a uniform sharkskin-like pattern. For shear rates higher than that, the contact angle decreased [192].

Due to their water repellency, superhydrophobic surfaces can be used as anti-icing surfaces [9, 193, 194]. A thin (127 μm), micro-nanostructured superhydrophobic sheet of Teflon fluoroethylene propylene (FEP) was fabricated using a laser micro-textured aluminum template by means of the hot embossing technique [194]. Teflon FEP was hot embossed at three different temperatures (290°C, 300°C and 310°C) under a constant pressure of 30 kPa for 5 min. Although increasing the temperature resulted in an opaque sheet, it also led to a rougher surface. The resultant water CA were >160° and the sliding angle <6°, with a reduction in ice detachment strength of 33% compared to a smooth non-textured Teflon FEP sheet [194].

Konstantinou et al. [11] optimized hot embossing parameters to produce thermoplastic microfluidic devices. Their main objective was to achieve a medium volume production having a high-quality replication. The optimized hot embossing parameters combined with improved milling processes and methods for epoxy mold fabrication produced microfluidic devices at a production rate of about 50 devices per week, which can be considered as an optimal medium volume production rate.

One of the main differences between the Wenzel and Cassie-Baxter models is the contact of the water droplet with the surface roughness.

In the Wenzel model, the water droplet is in contact with all points on the surface. Such enhanced interfacial contact between solid and liquid increases the hydrophobicity. However, in the Cassie-Baxter model, an interfacial solid-air-liquid composite is formed. These trapped air pockets hinder the water droplet from penetrating into the roughness. In this case, the height of the surface structures has a stronger influence. Taller microstructures favor a Cassie-Baxter regime, while shorter microstructures lead to a Wenzel regime [95].

Nayak et al. [95] fabricated superhydrophobic poly(dimethylsiloxane) (PDMS) using a direct replication technique on micro-nanostructured Si and Ti templates produced by an ultrafast laser. This approach resulted in a maximum CA greater than 154° and a sliding angle less than 6°. A threshold value θ_c, which obeys both Wenzel and Cassie-Baxter regimes, was obtained by equating both equations [95, 195].

$$\cos\theta_c = (f - 1)/(r - f) \tag{8}$$

This threshold angle is a function of r (the roughness factor in Wenzel's regime, which is larger than 1) and f (the dimensionless area fraction of the solid surface in contact with a liquid, which is smaller than 1). By linear curve fitting and solving two the empirical equations of the Wenzel and Cassie regimes, the threshold angle and the transition structure height were calculated as 129° and 4.2 μm, respectively. The latter value is the height necessary to pass from a Cassie-Baxter to a Wenzel regime [95], while the former value is the angle below which the Wenzel state is more practical energetically. However, a metastable Cassie-Baxter regime is also possible [195]. At angles greater than this threshold angle, only the Cassie-Baxter state is achievable. Note, however, that in addition to the structure height, the structures' shape, sizing and spacing also determine the CA.

In addition to these simple geometric shapes, more complicated geometries have been attempted such as simple-short, simple-long, stepped-short, and stepped-long nanopillar structures (Figure 13) [189]. Stepped-long nanopillars produced the most superhydrophobic surfaces. The other structures led to a WCA of <120°.

Figure 13. Schematic illustration of various types of nanopillars on anodic aluminum oxide (AAO) (first line) and imprinted polymer surfaces [189].

Using the *r* ratio in Wenzel's model and the *f* ratio in Cassie-Baxter's model, both recalculated based on the nanopillar geometries [189], the predicted WCA values of the various nanopillars types were compared with the experimental results. The resultant WCA values for the simple-short nanopillars matched predicted values from Wenzel's model. However, the observed WCA for the stepped-long nanopillars was consistent with the predicted value using Cassie-Baxter's model. The circular pillar diameter influences the CA of a microstructured PP surface produced by hot embossing [196]. Increased pillar diameter increases the CA, in agreement with Wenzel's theory. The highest CA (152°) was obtained at a diameter of 40 μm; pillars having a larger diameter produced a lower CA. This observation is based on a Cassie-Baxter model [196]. Micropillar height and the spacing between pillars also influence the CA.

5.3. Superhydrophobic Polymer Surfaces by Micro-Injection Molding

Cao et al. [197] fabricated superhydrophobic high-density polyethylene (HDPE) surfaces with water CA of around 152° using a nano-injection molding technique. The template was a porous anodic aluminum

(PAA) with pore diameters of 200 nm. The PE melts infiltrated into the nanopores under elevated temperature (160ºC), pressure generated by capillary forces and vibration (to help polymer chains to be more oriented while filling up the nanopores). A NaOH aqueous solution was used to remove the template from the solidified polymer. Although this technique eliminates some demolding issues, it may not always be applicable in regard to the template thickness, materials and conditions.

A couple of studies have assessed injection molded superhydrophobic polypropylene (PP) surfaces with microstructures and hierarchical anisotropic micro-nanostructures (dual structures) [62, 198]. Microstructured inserts were fabricated using a microworking robot on aluminum foil, and micro-nanostructured inserts were obtained by anodizing aluminum foil. Optimal results were achieved by adjusting the parameters of the microworking robot and the anodization process conditions. Although in some cases microstructures are sufficient to achieve a superhydrophobic surface, dual surfaces (i.e., surfaces with micro-nanostructures) always lead to a superhydrophobic surface with a CA greater than 150º.

Weng et al. [199], using a micro-injection molding process, produced superhydrophobic PP surfaces with a WCA of about 163º and sliding angle (SA) of 5º. Due to the low SA, the surface acted as a self-cleaning surface. High-aspect ratio PP and PC combined micro-nanostructured surfaces were produced using an AAO mold insert [51]. The combination of micro- and nanostructures resulted in a water CA of more than 150º. Moreover, higher mold temperature produced a higher water CA.

A practical application for textured superhydrophobic surfaces is microfluidic devices. Flow rate increases 186% in a PDMS channel having textured superhydrophobic inner walls [61]. A femtosecond laser textured a copper surface and the textures were then transferred to the PDMS surface. Among the multiple types of produced textures (i.e., laser-induced periodic surface structures (LIPSS), nano roughness, micro-grooves with nano-ripples, micro-grooves with nano-bumps, micro-bumps with nano-ripples, and micro-bumps with nano-bumps), micro-bumps with nano-bumps achieved the highest CA and lowest roll off angle for both the copper and PDMS surfaces. The wetting behavior of the metallic insert and the polymer replica obeyed the same trend with the different surface structures [61].

So far, this chapter has focused on hierarchically structured polymer surfaces fabricated on flat surfaces. However, a recent study has focused on the fabrication of rigid curved PP surfaces with accurately controlled hierarchical structures using a microworking robot (for making an aluminum foil template having a 0.2 mm thickness) and a micro-injection molding system [200]. Both convex and concave surfaces, as well as spherical and cylindrical structures, were manufactured. Interestingly, the highest static CA and the lowest CAH and SA were obtained from hierarchical micro-microstructured superhydrophobic polymer surfaces [195]. There are two main advantages in using micro-microstructures rather than micro-nanostructures: 1) a simple one-step fabrication process by a microrobot technique and 2) the durability and mechanical robustness of the micro-microstructured surfaces in comparison to micro-nanostructured surfaces.

A proper design of features could lead to a higher water CA. For cone-shape structures, as the base diameter of the individual cones is reduced, the distances between the cone tips are reduced and, as such, the CA increases [50]. However, the peak-to-peak distance should not decrease indiscriminately, as the formed compact surface would cause the drops to spread out more on the surface.

5.4. Durability of Superhydrophobic Micro-Nanostructured Polymer Surfaces

One of the main challenges for the production of superhydrophobic surfaces is the durability or mechanical resistance of the surface [201, 202]. The superhydrophobicity of PTFE surfaces fabricated by the hot embossing process remained stable after 120 days [25]. The thermal stability of these superhydrophobic PTFE surfaces was also confirmed for temperatures up to 340°C [25]. However, structure height decreases with abrasion or wear. Micro-injection molded superhydrophobic LSR surfaces subjected to abrasion had a water CA <150° in all test conditions. A superhydrophobic polyethylene (LDPE) surface remained dry after 5 h of immersion in water at a pressure of 55 kPa. Mechanical abrasion, from a gloved hand and a reciprocating abrasion machine at a pressure of 32 kPa, did not result in the CA being reduced below 150° even after 6000 abrasion cycles [190].

Huovinen et al. [203] introduced a new avenue for fabricating mechanically robust superhydrophobic polymer surfaces. Surface structures were protected against wear and compression forces by larger sacrificial micropillars, called protective pillars (Figure 14). Multistructures, micro-microstructures and micro-nanostructures were fabricated on aluminum foil using a microworking robot and replicated on polypropylene surfaces via micro-injection molding. The protective features then replaced some of the hierarchically structured pillars at a given replacement ratio.

The wetting behavior of superhydrophobic surfaces with protective shields showed remarkable robustness against abrasive wear and mechanical compression (1-5 N and 4-40 MPa, respectively) compared to unprotected structured surfaces.

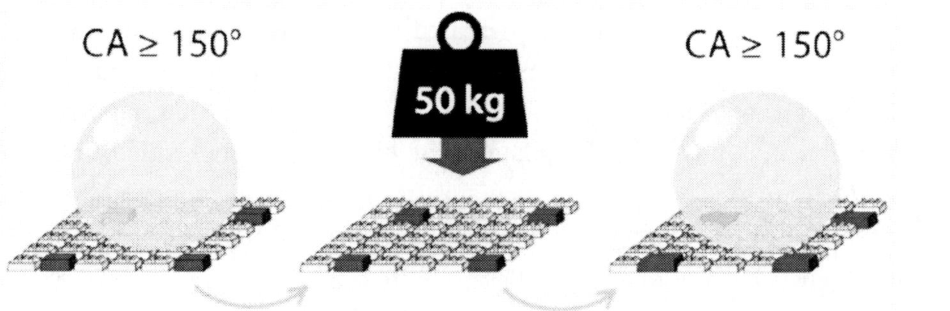

Figure 14. Schematic illustration of protective pillars for increasing the mechanical robustness of superhydrophobic polymer surfaces [203].

The aging and recovery processes of superhydrophobic silicone rubber surfaces have also been investigated [204]. Due to a lower Gibbs energy, the Wenzel regime is more stable than the Cassie regime. On the other hand, during the aging process, the presence of hydrophilic groups increases on the surface, and the surface roughness will increase the hydrophilicity. Therefore, the recovery process becomes difficult due to the large energy barrier between the Wenzel and Cassie regimes. The development of a Wenzel regime upon aging leads to incomplete recovery, meaning a large CAH even when the static contact angle is large [204]. The stability of the recovery process of superhydrophobic materials can be enhanced by reinforcing the surface chemical composition to delay the aging procedure and improving surface roughness to facilitate the Wenzel-Cassie transition [204]. The static and dynamic water CA of silicone rubber during the aging and recovery process were measured and were in good agreement with this theory [205]. The static CA of the superhydrophobic surface could be improved to a desirable degree (from 127° after aging to 148° after recovery with an original WCA of 151°). However, the dynamic CA had negligible recovery even after a long period [205].

Conclusion

The demand for micro-nanofeatured parts is increasing rapidly due to the numerous applications in various high-tech areas, including the production of superhydrophobic surfaces. Hot embossing and micro-injection molding offer the potential of high-quality replication and fast production to address this demand. This chapter addressed several elements of micro-injection molding and hot embossing of micro-nanostructured polymer surfaces in the fabrication (pre-processing, processing and post-processing (demolding)).

We reviewed the potential of various insert materials, particularly polymeric and hybrid inserts. Appropriate processing conditions are required to modify an insert's surface. High-quality replication depends on carefully controlling processing parameters to ensure optimal filling and demolding. The specific selected parameters depend on the employed polymer as well as the size and shape of both the mold and the micro-nanostructures. As the effects from the main parameters are not consistent depending on the given processing conditions, much consideration must be made for achieving the optimal results. The quality of the final product for both hot embossing and micro-injection molding is strongly influenced by mold temperature and pressure. Injection velocity, to a lesser degree, also affects the final product quality. Furthermore, as non-damaging demolding is essential for ensuring high-quality replication of micro-nanostructures, the capabilities of the various demolding methods must be carefully considered.

Superhydrophobic surfaces are achieved when micro-nanostructures are present on a surface. Both micro-injection molding and hot embossing techniques can produce these necessary structures. Successful use of both approaches relies on using a well-structured insert, carefully selecting the processing parameters and demolding without damage to the produced micro-nanostructures.

REFERENCES

[1] Jafari, R., Asadollahi, S., and Farzaneh, M., "Applications of plasma technology in development of superhydrophobic surfaces," *Plasma Chemistry and Plasma Processing,* vol. 33, no. 1, pp. 177-200, 2013.

[2] Stormonth-Darling, J. M., Pedersen, R. H., How, C., and Gadegaard, N., "Injection moulding of ultra high aspect ratio nanostructures using coated polymer tooling, " *Journal of Micromechanics and Microengineering,* vol. 24, no. 7, p. 075019, 2014.

[3] Momen, G., and Farzaneh, M., "Simple process to fabricate a superhydrophobic coating," *Micro & Nano Letters,* vol. 6, no. 6, pp. 405-407, 2011.

[4] Momen, G., and Farzaneh, M., "Superhydrophobic RTV silicone rubber coatings on anodized aluminium surfaces," in *Materials Science Forum*, 2012, vol. 706, pp. 2546-2551: Trans Tech Publ.

[5] Moon, S. d., Lee, N., and Kang, S., "Fabrication of a microlens array using micro-compression molding with an electroformed mold insert," *Journal of Micromechanics and Microengineering,* vol. 13, no. 1, p. 98, 2002.

[6] Jafari, R., and Farzaneh, M., "Development a simple method to create the superhydrophobic composite coatings," *Journal of Composite Materials,* vol. 47, no. 25, pp. 3125-3129, 2013.

[7] Momen, G., Farzaneh, M., and Jafari, R., "Wettability behaviour of RTV silicone rubber coated on nanostructured aluminium surface," *Applied Surface Science,* vol. 257, no. 15, pp. 6489-6493, 2011.

[8] Matschuk, M., and Larsen, N. B., "Injection molding of high aspect ratio sub-100 nm nanostructures," *Journal of Micromechanics and Microengineering,* vol. 23, no. 2, p. 025003, 2012.

[9] Momen, G., Jafari, R., and Farzaneh, M., "Ice repellency behaviour of superhydrophobic surfaces: Effects of atmospheric icing conditions and surface roughness," *Applied Surface Science,* vol. 349, pp. 211-218, 2015.

[10] Maghsoudi, K., Jafari, R., Momen, G., and Farzaneh, M., "Micro-nanostructured polymer surfaces using injection molding: A review," *Materials Today Communications,* vol. 13, pp. 126-143, 2017.

[11] Konstantinou, D., Shirazi, A., Sadri, A., and Young, E. W., "Combined hot embossing and milling for medium volume production of thermoplastic microfluidic devices," *Sensors and Actuators B: Chemical,* vol. 234, pp. 209-221, 2016.

[12] Park, S., Lee, W., Moon, S., Yoo, Y., and Cho, Y., "Injection molding micro patterns with high aspect ratio using a polymeric flexible stamper," *eXPRESS Polymer Letters,* vol. 5, no. 11, pp. 950-958, 2011.

[13] Matschuk, M., Bruus, H., and Larsen, N. B., "Nanostructures for all-polymer microfluidic systems," *Microelectronic Engineering,* vol. 87, no. 5, pp. 1379-1382, 2010.

[14] Hansen, H. N., Hocken, R., and Tosello, G., "Replication of micro and nano surface geometries," *CIRP Annals-Manufacturing Technology,* vol. 60, no. 2, pp. 695-714, 2011.

[15] Michaeli, W., Spennemann, A., and Gärtner, R., "New plastification concepts for micro injection moulding," *Microsystem Technologies,* vol. 8, no. 1, pp. 55-57, 2002.

[16] Greener, J., and Wimberger-Friedl, R., "*Precision Injection Molding*," Carl Hanser, Munich, 2006.

[17] Masato, D., Sorgato, M., and Lucchetta, G., "Analysis of the influence of part thickness on the replication of micro-structured surfaces by injection molding," *Materials & Design,* vol. 95, pp. 219-224, 2016.

[18] Liou, A. C., and Chen, R. H., "Injection molding of polymer micro- and sub-micron structures with high-aspect ratios," *The International Journal of Advanced Manufacturing Technology,* vol. 28, no. 11-12, pp. 1097-1103, 2006.

[19] Xie, L., Shen, L., and Jiang, B., "Modelling and simulation for micro injection molding process," in *Computational Fluid Dynamics Technologies and Applications*: InTech, 2011.

[20] Toosi, S. Falah, *"Superhydrophobic polymeric surfaces: fabrication, wettability, and antibbacterial activity,"* Doctoral dissertation, University of British Columbia, 2017.

[21] Wang, J., Yi, P., Deng, Y., Peng, L., Lai, X., and Ni, J., "Recovery behavior of thermoplastic polymers in micro hot embossing process," *Journal of Materials Processing Technology,* vol. 243, pp. 205-216, 2017.

[22] Sahli, M., Millot, C., Roques-Carmes, C., Malek, C. K., Barriere, T., and Gelin, J., "Quality assessment of polymer replication by hot embossing and micro-injection moulding processes using scanning mechanical microscopy," *Journal of Materials Processing Technology,* vol. 209, no. 18-19, pp. 5851-5861, 2009.

[23] Toosi, S. F., Moradi, S., Ebrahimi, M., and Hatzikiriakos, S. G., "Microfabrication of polymeric surfaces with extreme wettability

using hot embossing," *Applied Surface Science,* vol. 378, pp. 426-434, 2016.

[24] Jucius, D., et al., "Hot embossing of PTFE: towards superhydrophobic surfaces," *Applied Surface Science,* vol. 257, no. 6, pp. 2353-2360, 2011.

[25] Gong, D., Long, J., Fan, P., Jiang, D., Zhang, H., and Zhong, M., "Thermal stability of micro–nano structures and superhydrophobicity of polytetrafluoroethylene films formed by hot embossing via a picosecond laser ablated template," *Applied Surface Science,* vol. 331, pp. 437-443, 2015.

[26] Malek, C. K., Coudevylle, J. R., Jeannot, J. C., and Duffait, R., "Revisiting micro hot-embossing with moulds in non-conventional materials," *Microsystem Technologies,* vol. 13, no. 5-6, pp. 475-481, 2007.

[27] Peng, L., Deng, Y., Yi, P., and Lai, X., "Micro hot embossing of thermoplastic polymers: a review," *Journal of Micromechanics and Microengineering,* vol. 24, no. 1, p. 013001, 2013.

[28] Heckele, M., Bacher, W., and Müller, K., "Hot embossing-the molding technique for plastic microstructures," *Microsystem Technologies,* vol. 4, no. 3, pp. 122-124, 1998.

[29] Guo, Y., Liu, G., Zhu, X., and Tian, Y., "Analysis of the demolding forces during hot embossing," *Microsystem Technologies,* vol. 13, no. 5-6, pp. 411-415, 2007.

[30] He, P., et al., "Bulk metallic glass mold for high volume fabrication of micro optics," *Microsystem Technologies,* vol. 22, no. 3, pp. 617-623, 2016.

[31] Packianather, M., Griffiths, C., and Kadir, W., "Micro injection moulding process parameter tuning," *Procedia CIRP,* vol. 33, pp. 400-405, 2015.

[32] Huang, C., "Polymeric nanofeatures of 100 nm using injection moulding for replication," *Journal of Micromechanics and Microengineering,* vol. 17, no. 8, p. 1518, 2007.

[33] Macintyre, D., and Thoms, S., "The fabrication of high resolution features by mould injection," *Microelectronic Engineering,* vol. 41, pp. 211-214, 1998.

[34] Pakkanen, T. T., Hietala, J., Pääkkönen, E. J., Pääkkönen, P., Jääskeläinen, T., and Kaikuranta, T., "Replication of sub-micron features using amorphous thermoplastics," *Polymer Engineering & Science,* vol. 42, no. 7, pp. 1600-1608, 2002.

[35] Schift, H. et al., "Nanoreplication in polymers using hot embossing and injection molding," *Microelectronic Engineering,* vol. 53, no. 1-4, pp. 171-174, 2000.

[36] Gadegaard, N., Mosler, S., and Larsen, N. B., "Biomimetic polymer nanostructures by injection molding," *Macromolecular Materials and Engineering,* vol. 288, no. 1, pp. 76-83, 2003.

[37] Yoshii, M., Kuramoto, H., and Kato, K., "Experimental study of transcription of minute width grooves in injection molding," *Polymer Engineering & Science,* vol. 34, no. 15, pp. 1211-1218, 1994.

[38] Holthusen, A. K., Riemer, O., Schmütz, J., and Meier, A., "Mold machining and injection molding of diffractive microstructures," *Journal of Manufacturing Processes,* vol. 26, pp. 290-294, 2017.

[39] Griffiths, C., Dimov, S., Scholz, S., Tosello, G., and Rees, A., "Influence of injection and cavity pressure on the Demoulding force in micro-injection moulding," *Journal of Manufacturing Science and Engineering,* vol. 136, no. 3, p. 031014, 2014.

[40] Heyderman, L., Schift, H., David, C., Gobrecht, J., and Schweizer, T., "Flow behaviour of thin polymer films used for hot embossing lithography," *Microelectronic Engineering,* vol. 54, no. 3-4, pp. 229-245, 2000.

[41] Bellantone, V., Surace, R., Trotta, G., and Fassi, I., "Replication capability of micro injection moulding process for polymeric parts manufacturing," *The International Journal of Advanced Manufacturing Technology,* vol. 67, no. 5-8, pp. 1407-1421, 2013.

[42] Feldmann, M., "The effects of the injection moulding temperature on the mechanical properties and morphology of polypropylene man-

made cellulose fibre composites," *Composites Part A: Applied Science and Manufacturing,* vol. 87, pp. 146-152, 2016.

[43] Cunha, L., et al., "Performance of chromium nitride and titanium nitride coatings during plastic injection moulding," *Surface and Coatings Technology,* vol. 153, no. 2, pp. 160-165, 2002.

[44] Zhang, Y., Hansen, H. N., and Sørensen, S., "Replication of micro-pillars by PEEK injection moulding with CrN-coated Ni tool," *The International Journal of Advanced Manufacturing Technology,* vol. 80, no. 1-4, pp. 383-388, 2015.

[45] Griffiths, C. et al., "Investigation of polymer inserts as prototyping tooling for micro injection moulding," *The International Journal of Advanced Manufacturing Technology,* vol. 47, no. 1-4, pp. 111-123, 2010.

[46] Yoon, S. h., Srirojpinyo, C., Lee, J. S., Mead, J. L., Matsui, S., and Barry, C. M., "Evaluation of novel tooling for nanoscale injection molding," in *Smart Structures and Materials*, International Society for Optics and Photonics, pp. 107-116, 2005.

[47] Menotti, S., Hansen, H. N., Bissacco, G., Calaon, M., Tang, P. T., and Ravn, C., "Injection molding of nanopatterned surfaces in the sub-micrometer range with induction heating aid," *The International Journal of Advanced Manufacturing Technology,* vol. 74, no. 5-8, pp. 907-916, 2014.

[48] Kimura, F., Kadoya, S., and Kajihara, Y., "Effects of molding conditions on injection molded direct joining using a metal with nano-structured surface," *Precision Engineering,* vol. 45, pp. 203-208, 2016.

[49] Shen, Y., and Wu, W., "An analysis of the three-dimensional micro-injection molding," *International Communications in Heat and Mass Transfer,* vol. 29, no. 3, pp. 423-431, 2002.

[50] Hopmann, C., Behmenburg, C., Recht, U., and Zeuner, K., "Injection molding of superhydrophobic liquid silicone rubber surfaces," *Silicon,* vol. 6, no. 1, pp. 35-43, 2014.

[51] Zhou, M., Xiong, X., Jiang, B., and Weng, C., "Fabrication of high aspect ratio nanopillars and micro/nano combined structures with

hydrophobic surface characteristics by injection molding," *Applied Surface Science,* vol. 427, pp. 854-860, 2018.

[52] Celia, E., Darmanin, T., de Givenchy, E. T., Amigoni, S., and Guittard, F., "Recent advances in designing superhydrophobic surfaces," *Journal of Colloid and Interface Science,* vol. 402, pp. 1-18, 2013.

[53] Stormonth-Darling, J. M., and Gadegaard, N., "Injection moulding difficult nanopatterns with hybrid polymer inlays," *Macromolecular Materials and Engineering,* vol. 297, no. 11, pp. 1075-1080, 2012.

[54] Søgaard, E., "*Injection molded self-cleaning surfaces,*" Doctoral dissertation, DTU Nanotech, 2014.

[55] Stormonth-Darling, J. M., Saeed, A., Reynolds, P. M., and Gadegaard, N., "Injection Molding Micro-and Nanostructures in Thermoplastic Elastomers," *Macromolecular Materials and Engineering,* vol. 301, no. 8, pp. 964-971, 2016.

[56] Su, Y. C., Shah, J., and Lin, L., "Implementation and analysis of polymeric microstructure replication by micro injection molding," *Journal of Micromechanics and Microengineering,* vol. 14, no. 3, p. 415, 2003.

[57] Xie, L., Niesel, T., Leester-Schädel, M., Ziegmann, G., and Büttgenbach, S., "A novel approach to realize the local precise variotherm process in micro injection molding," *Microsystem Technologies,* vol. 19, no. 7, pp. 1017-1023, 2013.

[58] Zhang, N., Srivastava, A. P., Browne, D. J., and Gilchrist, M. D., "Performance of nickel and bulk metallic glass as tool inserts for the microinjection molding of polymeric microfluidic devices," *Journal of Materials Processing Technology,* vol. 231, pp. 288-300, 2016.

[59] Zhang, N., et al., "Manufacturing microstructured tool inserts for the production of polymeric microfluidic devices," *Journal of Micromechanics and Microengineering,* vol. 25, no. 9, p. 095005, 2015.

[60] Zhang, N., Chu, J., Byrne, C. J., Browne, D., and Gilchrist, M., "Replication of micro/nano-scale features by micro injection molding with a bulk metallic glass mold insert," *Journal of*

Micromechanics and Microengineering, vol. 22, no. 6, p. 065019, 2012.

[61] Sarbada, S., and Shin, Y. C., "Superhydrophobic contoured surfaces created on metal and polymer using a femtosecond laser," *Applied Surface Science,* vol. 405, pp. 465-475, 2017.

[62] Puukilainen, E., Rasilainen, T., Suvanto, M., and Pakkanen, T. A., "Superhydrophobic polyolefin surfaces: controlled micro-and nanostructures," *Langmuir,* vol. 23, no. 13, pp. 7263-7268, 2007.

[63] Kim, S., Jeong, J., and Youn, J., "Nanopattern insert molding," *Nanotechnology,* vol. 21, no. 20, p. 205302, 2010.

[64] Kim, S. H., Sul, I. H., Jeong, J. h., Song, Y. S., and Youn, J. R., "Robust fabrication and evaluation of nanopattern insert molded parts," *European Polymer Journal,* vol. 49, no. 6, pp. 1437-1445, 2013.

[65] Yang, C., Yin, X. H., and Cheng, G. M., "Microinjection molding of microsystem components: new aspects in improving performance," *Journal of Micromechanics and Microengineering,* vol. 23, no. 9, p. 093001, 2013.

[66] Rogers, J. A. and Lee, H. H., *Unconventional nanopatterning techniques and applications*. John Wiley & Sons, 2008.

[67] Pfleging, W., Hanemann, T., Torge, M., and Bernauer, W., "Rapid fabrication and replication of metal, ceramic and plastic mould inserts for application in microsystem Technologies," *Proceedings of the Institution of Mechanical Engineers, Part C: Journal of Mechanical Engineering Science,* vol. 217, no. 1, pp. 53-63, 2003.

[68] Zhong, Z., Ng, H., Chen, S., and Shan, X., "Hot roller embossing of multi-dimensional microstructures using elastomeric molds," *Microsystem Technologies,* vol. 24, no. 3, pp. 1443-1452, 2018.

[69] Lee, B. K. and Lee, B. y., "Investigation of thermoplastic hot embossing process using soft polydimethylsiloxane (PDMS) micromold," *Journal of Mechanical Science and Technology,* vol. 29, no. 12, pp. 5063-5067, 2015.

[70] Liu, Y., et al., "Polymeric microlens array fabricated with PDMS mold-based hot embossing," *Journal of Micromechanics and Microengineering,* vol. 24, no. 9, p. 095028, 2014.

[71] Chen, J., Gu, C., Lin, H., and Chen, S. C., "Soft mold-based hot embossing process for precision imprinting of optical components on non-planar surfaces," *Optics Express,* vol. 23, no. 16, pp. 20977-20985, 2015.

[72] Sahli, M., Millot, C., Gelin, J. C., and Barrière, T., "The manufacturing and replication of microfluidic mould inserts by the hot embossing process," *Journal of Materials Processing Technology,* vol. 213, no. 6, pp. 913-925, 2013.

[73] Kim, M., Moon, B. U., and Hidrovo, C. H., "Enhancement of the thermo-mechanical properties of PDMS molds for the hot embossing of PMMA microfluidic devices," *Journal of Micromechanics and Microengineering,* vol. 23, no. 9, p. 095024, 2013.

[74] Xie, P., et al., "Rapid hot embossing of polymer microstructures using carbide-bonded graphene coating on silicon stampers," *Surface and Coatings Technology,* vol. 258, pp. 174-180, 2014.

[75] Despa, M., Kelly, K., and Collier, J., "Injection molding of polymeric LIGA HARMs," *Microsystem Technologies,* vol. 6, no. 2, pp. 60-66, 1999.

[76] Lin, H. Y., and Young, W. B., "Analysis of the filling capability to the microstructures in micro-injection molding," *Applied Mathematical Modelling,* vol. 33, no. 9, pp. 3746-3755, 2009.

[77] Lin, H. Y., Chang, C. H., and Young, W. B., "Experimental study on the filling of nano structures with infrared mold surface heating," *International Polymer Processing,* vol. 26, no. 1, pp. 73-81, 2011.

[78] Piotter, V., Mueller, K., Plewa, K., Ruprecht, R., and Hausselt, J., "Performance and simulation of thermoplastic micro injection molding," *Microsystem Technologies,* vol. 8, no. 6, pp. 387-390, 2002.

[79] Giboz, J., Copponnex, T., and Mélé, P., "Microinjection molding of thermoplastic polymers: a review," *Journal of Micromechanics and Microengineering,* vol. 17, no. 6, p. R96, 2007.

[80] Rötting, O., Röpke, W., Becker, H., and Gärtner, C., "Polymer microfabrication technologies," *Microsystem Technologies,* vol. 8, no. 1, pp. 32-36, 2002.

[81] Scholz, S., Griffiths, C., Dimov, S., Brousseau, E., Lalev, G., and Petkov, P., "Manufacturing routes for replicating micro and nano surface structures with bio-mimetic applications," *CIRP Journal of Manufacturing Science and Technology,* vol. 4, no. 4, pp. 347-356, 2011.

[82] Becker, H., and Gärtner, C., "Polymer microfabrication technologies for microfluidic systems," *Analytical and Bioanalytical Chemistry,* vol. 390, no. 1, pp. 89-111, 2008.

[83] Wang, Q. L., Zhu, D. D., Chen, Y., and Guo, X. D., "A fabrication method of microneedle molds with controlled microstructures," *Materials Science and Engineering: C,* vol. 65, pp. 135-142, 2016.

[84] Schift, H., Urwyler, P., and Kristiansen, P. M., "Surface-patterned micromechanical elements by polymer injection molding with hybrid molds," *Journal of Vacuum Science & Technology B, Nanotechnology and Microelectronics: Materials, Processing, Measurement, and Phenomena,* vol. 31, no. 6, p. 06FD01, 2013.

[85] Michaeli, W., and Kamps, T., "Micro assembly injection moulding with plasma treated inserts," *Microsystem Technologies,* vol. 14, no. 12, pp. 1903-1907, 2008.

[86] Pedersen, R. H., Xu, Q., Stormonth-Darling, J. M., and Gadegaard, N., "Strategies for high quality injection moulding of polymer nanopillars," *Macromolecular Materials and Engineering,* vol. 300, no. 2, pp. 172-180, 2015.

[87] Ting, C. J., Huang, M. C., Tsai, H. Y., Chou, C. P., and Fu, C. C., "Low cost fabrication of the large-area anti-reflection films from polymer by nanoimprint/hot-embossing technology," *Nanotechnology,* vol. 19, no. 20, p. 205301, 2008.

[88] Hansen, T. S., Selmeczi, D., and Larsen, N. B., "Fast prototyping of injection molded polymer microfluidic chips," *Journal of Micromechanics and Microengineering,* vol. 20, no. 1, p. 015020, 2009.

[89] Uhlmann, E., Mullany, B., Biermann, D., Rajurkar, K., Hausotte, T., and Brinksmeier, E., "Process chains for high-precision components with micro-scale features," *CIRP Annals-Manufacturing Technology,* vol. 65, no. 2, pp. 549-572, 2016.

[90] Nayak, B. K., and Gupta, M. C., "Self-organized micro/nano structures in metal surfaces by ultrafast laser irradiation," *Optics and Lasers in Engineering,* vol. 48, no. 10, pp. 940-949, 2010.

[91] Chun, D. M., Davaasuren, G., Ngo, C. V., Kim, C. S., Lee, G. Y., and Ahn, S. H., "Fabrication of transparent superhydrophobic surface on thermoplastic polymer using laser beam machining and compression molding for mass production," *CIRP Annals-Manufacturing Technology,* vol. 63, no. 1, pp. 525-528, 2014.

[92] Malek, C. G. K., "Laser processing for bio-microfluidics applications (part I)," *Analytical and Bioanalytical Chemistry,* vol. 385, no. 8, pp. 1351-1361, 2006.

[93] Malek, C. G. K., "Laser processing for bio-microfluidics applications (part II)," *Analytical and Bioanalytical Chemistry,* vol. 385, no. 8, pp. 1362-1369, 2006.

[94] Ahmmed, K. T., Ling, E. J. Y., Servio, P., and Kietzig, A. M., "Introducing a new optimization tool for femtosecond laser-induced surface texturing on titanium, stainless steel, aluminum and copper," *Optics and Lasers in Engineering,* vol. 66, pp. 258-268, 2015.

[95] Nayak, B. K., Caffrey, P. O., Speck, C. R., and Gupta, M. C., "Superhydrophobic surfaces by replication of micro/nano-structures fabricated by ultrafast-laser-microtexturing," *Applied Surface Science,* vol. 266, pp. 27-32, 2013.

[96] Lehr, J., and Kietzig, A. M., "Production of homogenous micro-structures by femtosecond laser micro-machining," *Optics and Lasers in Engineering,* vol. 57, pp. 121-129, 2014.

[97] Milionis, A. et al., "Superhydrophobic/superoleophilic magnetic elastomers by laser ablation," *Applied Surface Science,* vol. 351, pp. 74-82, 2015.

[98] Du, Q., Ai, J., Qin, Z., Liu, J., and Zeng, X., "Fabrication of superhydrophobic/superhydrophilic patterns on polyimide surface by

ultraviolet laser direct texturing," *Journal of Materials Processing Technology,* vol. 251, pp. 188-196, 2018.

[99] Kong, J., Xu, Y., Yung, K. L., Xie, Y., and He, L., "Enhanced polymer melts flow though nanoscale channels under vibration," *The Journal of Physical Chemistry C,* vol. 113, no. 2, pp. 624-629, 2008.

[100] Kim, E., Xia, Y., and Whitesides, G. M., "Micromolding in capillaries: applications in materials science," *Journal of the American Chemical Society,* vol. 118, no. 24, pp. 5722-5731, 1996.

[101] Young, W. B., "Analysis of filling distance in cylindrical microfeatures for microinjection molding," *Applied Mathematical Modelling,* vol. 31, no. 9, pp. 1798-1806, 2007.

[102] Song, M., Zhao, H., Liu, J., Liu, C., and Li, J., "Replication of large scale micro pillar array with different diameters by micro injection molding," *Microsystem Technologies,* pp. 1-10, 2016.

[103] Pranov, H., Rasmussen, H. K., Larsen, N. B., and Gadegaard, N., "On the injection molding of nanostructured polymer surfaces," *Polymer Engineering & Science,* vol. 46, no. 2, pp. 160-171, 2006.

[104] Yoo, Y. et al., "Injection molding of a nanostructured plate and measurement of its surface properties," *Current Applied Physics,* vol. 9, no. 2, pp. e12-e18, 2009.

[105] Piotter, V., Holstein, N., Plewa, K., Ruprecht, R., and Hausselt, J., "Replication of micro components by different variants of injection molding," *Microsystem Technologies,* vol. 10, no. 6-7, pp. 547-551, 2004.

[106] Smith, D. E., Tortorelli, D. A., and Tucker, C. L., "Analysis and sensitivity analysis for polymer injection and compression molding," *Computer Methods in Applied Mechanics and Engineering,* vol. 167, no. 3-4, pp. 325-344, 1998.

[107] Xu, B., Ooi, K., Wong, T., and Liu, C., "Study on the viscosity of the liquid flowing in microgeometry," *Journal of Micromechanics and Microengineering,* vol. 9, no. 4, p. 377, 1999.

[108] Rosenbaum, E. E., and Hatzikiriakos, S. G., "Wall slip in the capillary flow of molten polymers subject to viscous heating," *AIChE journal,* vol. 43, no. 3, pp. 598-608, 1997.

[109] Pirskanen, J., et al., "Replication of sub-micrometre features using microsystems Technology," *Plastics, Rubber and Composites,* vol. 34, no. 5-6, pp. 222-226, 2005.

[110] Zhou, M., Jiang, B., Weng, C., and Zhang, L., "Experimental study on the replication quality of micro–nano cross-shaped structure arrays in injection molding," *Microsystem Technologies,* vol. 23, no. 4, pp. 983-989, 2017.

[111] Yang, C., Li, L., Huang, H., Castro, J. M., and Yi, A. Y., "Replication characterization of microribs fabricated by combining ultraprecision machining and microinjection molding," *Polymer Engineering & Science,* vol. 50, no. 10, pp. 2021-2030, 2010.

[112] Sha, B., Dimov, S., Griffiths, C., and Packianather, M. S., "Micro-injection moulding: factors affecting the achievable aspect ratios," *The International Journal of Advanced Manufacturing Technology,* vol. 33, no. 1-2, pp. 147-156, 2007.

[113] Miikkulainen, V. et al., "Thin films of MoN, WN, and perfluorinated silane deposited from dimethylamido precursors as contamination resistant coatings on micro-injection mold inserts," *Surface and Coatings Technology,* vol. 202, no. 21, pp. 5103-5109, 2008.

[114] Saha, B., Toh, W. Q., Liu, E., Tor, S. B., Hardt, D. E., and Lee, J., "A review on the importance of surface coating of micro/nano-mold in micro/nano-molding processes," *Journal of Micromechanics and Microengineering,* vol. 26, no. 1, p. 013002, 2015.

[115] Michaeli, W., Klaiber, F., and Scholz, S., "Investigations in variothermal injection moulding of microstructures and microstructured surfaces," *Multi-Material Micro Manufacture,* 2008.

[116] Lucchetta, G., Sorgato, M., Carmignato, S., and Savio, E., "Investigating the technological limits of micro-injection molding in replicating high aspect ratio micro-structured surfaces," *CIRP Annals-Manufacturing Technology,* vol. 63, no. 1, pp. 521-524, 2014.

[117] Lucchetta, G., Ferraris, E., Tristo, G., and Reynaerts, D., "Influence of mould thermal properties on the replication of micro parts via injection moulding," *Procedia CIRP,* vol. 2, pp. 113-117, 2012.

[118] Griffiths, C., Tosello, G., Dimov, S., Scholz, S., Rees, A., and Whiteside, B., "Characterisation of demoulding parameters in micro-injection moulding," *Microsystem Technologies,* vol. 21, no. 8, pp. 1677-1690, 2015.

[119] Yokoi, H., Han, X., Takahashi, T., and Kim, W., "Effects of molding conditions on transcription molding of microscale prism patterns using ultra-high-speed injection molding," *Polymer Engineering & Science,* vol. 46, no. 9, pp. 1140-1146, 2006.

[120] Whiteside, B., Martyn, M., Coates, P., Allan, P., Hornsby, P., and Greenway, G., "Micromoulding: process characteristics and product properties," *Plastics, Rubber and Composites,* vol. 32, no. 6, pp. 231-239, 2003.

[121] Whiteside, B. R., Martyn, M. T., Coates, P. D., Greenway, G., Allen, P., and Hornsby, P., "Micromoulding: process measurements, product morphology and properties," *Plastics, Rubber and Composites,* vol. 33, no. 1, pp. 11-17, 2004.

[122] Zhao, J., Chen, G., Chan, P., and Debowski, M., "Investigation of interactions of polymer material, mould design and process condition in the micro moulding process," *SIMTech Technical Report,* Citeseer 2003.

[123] Wlodarski, P., and Pittman, J., "Replication of surface micro-features using variothermal injection molding: Application to micro-fluidics," *Polymer Engineering & Science.*

[124] Speranza, V., Liparoti, S., Calaon, M., Tosello, G., Pantani, R., and G. Titomanlio, "Replication of micro and nano-features on iPP by injection molding with fast cavity surface temperature evolution," *Materials & Design,* vol. 133, pp. 559-569, 2017.

[125] Meister, S., and Drummer, D., "Micro assembly injection moulding with variothermal mould tempering," *Microsystem Technologies,* vol. 23, no. 4, pp. 1017-1025, 2017.

[126] Baruffi, F., Calaon, M., and Tosello, G., "Effects of micro-injection moulding process parameters on accuracy and precision of thermoplastic elastomer micro rings," *Precision Engineering,* vol. 51, pp. 353-361, 2018.

[127] Lan, S., et al., "A parameter study on the micro hot-embossing process of glassy polymer for pattern replication," *Microelectronic Engineering,* vol. 86, no. 12, pp. 2369-2374, 2009.

[128] Su, Q., Zhang, N., and Gilchrist, M. D., "The use of variotherm systems for microinjection molding," *Journal of Applied Polymer Science,* vol. 133, no. 9, 2016.

[129] Wang, G., Zhao, G., Li, H., and Guan, Y., "Research on a new variotherm injection molding technology and its application on the molding of a large LCD panel," *Polymer-Plastics Technology and Engineering,* vol. 48, no. 7, pp. 671-681, 2009.

[130] Chen, M., Yao, D., and Kim, B., "Eliminating flow induced birefringence and minimizing thermally induced residual stresses in injection molded parts," *Polymer-Plastics Technology and Engineering,* vol. 40, no. 4, pp. 491-503, 2001.

[131] Michaeli, W., Opfermann, D., and Kamps, T., "Advances in micro assembly injection moulding for use in medical systems," *The International Journal of Advanced Manufacturing Technology,* vol. 33, no. 1-2, pp. 206-211, 2007.

[132] Sha, B., Dimov, S., Griffiths, C., and Packianather, M., "Investigation of micro-injection moulding: factors affecting the replication quality," *Journal of Materials Processing Technology,* vol. 183, no. 2, pp. 284-296, 2007.

[133] Chen, C. P., Chuang, M. T., Hsiao, Y. H., Yang, Y. K., and Tsai, C. H., "Simulation and experimental study in determining injection molding process parameters for thin-shell plastic parts via design of experiments analysis," *Expert Systems with Applications,* vol. 36, no. 7, pp. 10752-10759, 2009.

[134] Nakao, M., Tsuchiya, K., Sadamitsu, T., Ichikohara, Y., Ohba, T., and Ooi, T., "Heat transfer in injection molding for reproduction of sub-micron-sized features," *The International Journal of Advanced Manufacturing Technology,* vol. 38, no. 3, pp. 426-432, 2008.

[135] Song, M., Liu, Z., Wang, M., Yu, T., and Zhao, D., "Research on effects of injection process parameters on the molding process for

ultra-thin wall plastic parts," *Journal of Materials Processing Technology,* vol. 187, pp. 668-671, 2007.

[136] Ishizawa, N., Idei, K., Kimura, T., Noda, D., and Hattori, T., "Resin micromachining by roller hot embossing," *Microsystem Technologies,* vol. 14, no. 9-11, pp. 1381-1388, 2008.

[137] Zhao, J., Mayes, R., Chen, G., Chan, P., and Xiong, Z., "Polymer micromould design and micromoulding process," *Plastics, Rubber and Composites,* vol. 32, no. 6, pp. 240-247, 2003.

[138] Yu, L., Koh, C. G., Lee, L. J., Koelling, K. W., and Madou, M. J., "Experimental investigation and numerical simulation of injection molding with micro-features," *Polymer Engineering & Science,* vol. 42, no. 5, pp. 871-888, 2002.

[139] Yoon, S. H., et al., "Effect of processing parameters, antistiction coatings, and polymer type when injection molding microfeatures," *Polymer Engineering and Science,* vol. 50, no. 2, p. 411, 2010.

[140] Sha, B., Dimov, S., Pham, D., and Griffiths, C., "Study of factors affecting aspect ratios achievable in micro-injection moulding," in *4M2005 Conference on Multi-material Micro Manufacture, Karlsrhue*, 2005.

[141] Surace, R., Trotta, G., Fassi, I., and Bellantone, V., *The micro injection moulding process for polymeric components manufacturing*. INTECH Open Access Publisher, 2012.

[142] Yang, C., Huang, H. X., Castro, J. M., and Yi, A. Y., "Replication characterization in injection molding of microfeatures with high aspect ratio: influence of layout and shape factor," *Polymer Engineering & Science,* vol. 51, no. 5, pp. 959-968, 2011.

[143] Eringen, A. C., and Okada, K., "A lubrication theory for fluids with microstructure," *International Journal of Engineering Science,* vol. 33, no. 15, pp. 2297-2308, 1995.

[144] Zhang, K., and Lu, Z., "Analysis of morphology and performance of PP microstructures manufactured by micro injection molding," *Microsystem Technologies,* vol. 14, no. 2, pp. 209-214, 2008.

[145] Ruprecht, R., Gietzelt, T., Müller, K., Piotter, V., and Haußelt, J., "Injection molding of microstructured components from plastics,

metals and ceramics," *Microsystem Technologies,* vol. 8, no. 4-5, pp. 351-358, 2002.

[146] Heckele, M., and Schomburg, W., "Review on micro molding of thermoplastic polymers," *Journal of Micromechanics and Microengineering,* vol. 14, no. 3, p. R1, 2003.

[147] Gornik, C., "Injection moulding of parts with microstructured surfaces for medical applications," in *Macromolecular Symposia,* 2004, vol. 217, no. 1, pp. 365-374: Wiley Online Library.

[148] Yoon, S. H., Padmanabha, P., Cha, N. G., Mead, J., and Barry, C., "Evaluation of vacuum venting for micro-injection molding," *International Polymer Processing,* vol. 26, no. 4, pp. 346-353, 2011.

[149] Sorgato, M., Babenko, M., Lucchetta, G., and Whiteside, B., "Investigation of the influence of vacuum venting on mould surface temperature in micro injection moulding," *The International Journal of Advanced Manufacturing Technology,* vol. 88, no. 1-4, pp. 547-555, 2017.

[150] Zhao, J., Mayes, R., Chen, G., Xie, H., and Chan, P. S., "Effects of process parameters on the micro molding process," *Polymer Engineering & Science,* vol. 43, no. 9, pp. 1542-1554, 2003.

[151] Su, Q. and Gilchrist, M. D., "Demolding forces for micron-sized features during micro-injection molding," *Polymer Engineering & Science,* vol. 56, no. 7, pp. 810-816, 2016.

[152] Delaney, K., Bissacco, G., and Kennedy, D., "A structured review and classification of demolding issues and proven solutions," *International Polymer Processing,* vol. 27, no. 1, pp. 77-90, 2012.

[153] Berger, G. R., Steffel, C., and Friesenbichler, W., "A study on the role of wetting parameters on friction in injection moulding," *International Journal of Materials and Product Technology,* vol. 52, no. 1-2, pp. 193-211, 2016.

[154] Wu, C. H., and Liang, W. J., "Effects of geometry and injection-molding parameters on weld-line strength," *Polymer Engineering & Science,* vol. 45, no. 7, pp. 1021-1030, 2005.

[155] Peng, Z., Gang, L., Yangchao, T., and Xuehong, T., "The properties of demoulding of Ni and Ni-PTFE moulding inserts," *Sensors and Actuators A: Physical,* vol. 118, no. 2, pp. 338-341, 2005.

[156] Sorgato, M., Masato, D., and Lucchetta, G., "Effects of machined cavity texture on ejection force in micro injection molding," *Precision Engineering,* vol. 50, pp. 440-448, 2017.

[157] Pontes, A., and Pouzada, A., "Ejection force in tubular injection moldings. Part I: Effect of processing conditions," *Polymer Engineering & Science,* vol. 44, no. 5, pp. 891-897, 2004.

[158] Pontes, A., Pouzada, A., Pantani, R., and Titomanlio, G., "Ejection force of tubular injection moldings. Part II: A prediction model," *Polymer Engineering & Science,* vol. 45, no. 3, pp. 325-332, 2005.

[159] Sasaki, T., Koga, N., Shirai, K., Kobayashi, Y., and Toyoshima, A., "An experimental study on ejection forces of injection molding," *Precision Engineering,* vol. 24, no. 3, pp. 270-273, 2000.

[160] Correia, M. S., Miranda, A. S., Oliveira, M. C., Capela, C. A., and Pouzada, A. S., "Analysis of friction in the ejection of thermoplastic mouldings," *The International Journal of Advanced Manufacturing Technology,* vol. 59, no. 9, pp. 977-986, 2012.

[161] Omar, F., Brousseau, E., Elkaseer, A., Kolew, A., Prokopovich, P., and Dimov, S., "Development and experimental validation of an analytical model to predict the demoulding force in hot embossing," *Journal of Micromechanics and Microengineering,* vol. 24, no. 5, p. 055007, 2014.

[162] Masato, D., Sorgato, M., and Lucchetta, G., "Effect of ultrasound vibration on the ejection friction in microinjection molding," *The International Journal of Advanced Manufacturing Technology,* pp. 1-14, 2018.

[163] Birkar, S., Mendible, G., Park, J. G., Mead, J. L., Johnston, S. P., and Barry, C. M., "Effect of feature spacing when injection molding parts with microstructured surfaces," *Polymer Engineering & Science,* vol. 56, no. 12, pp. 1330-1338, 2016.

[164] Beck, M., et al., "Improving stamps for 10 nm level wafer scale nanoimprint lithography," *Microelectronic Engineering,* vol. 61, pp. 441-448, 2002.

[165] Chou, S. Y., Krauss, P. R., and Renstrom, P. J., "Nanoimprint lithography," *Journal of Vacuum Science & Technology B: Microelectronics and Nanometer Structures Processing, Measurement, and Phenomena,* vol. 14, no. 6, pp. 4129-4133, 1996.

[166] Christiansen, A. B., Clausen, J. S., Mortensen, N. A., and Kristensen, A., "Injection moulding antireflective nanostructures," *Microelectronic Engineering,* vol. 121, pp. 47-50, 2014.

[167] Yoon, S. H., et al., "Enhancement of surface replication by gas assisted microinjection moulding," *Plastics, Rubber and Composites,* vol. 39, no. 7, pp. 293-299, 2010.

[168] Schift, H., Saxer, S., Park, S., Padeste, C., Pieles, U., and Gobrecht, J., "Controlled co-evaporation of silanes for nanoimprint stamps," *Nanotechnology,* vol. 16, no. 5, p. S171, 2005.

[169] Bhagat, S. D., and Gupta, M. C., "Superhydrophobic microtextured polycarbonate surfaces," *Surface and Coatings Technology,* vol. 270, pp. 117-122, 2015.

[170] Chakradhar, R., Kumar, V. D., Rao, J., and Basu, B. J., "Fabrication of superhydrophobic surfaces based on ZnO–PDMS nanocomposite coatings and study of its wetting behaviour," *Applied Surface Science,* vol. 257, no. 20, pp. 8569-8575, 2011.

[171] Brassard, J. D., Laforte, J. L., Blackburn, C., Perron, J., and Sarkar, D., "Silicone based superhydrophobic coating efficient to reduce ice adhesion and accumulation on aluminum under offshore arctic conditions," *Ocean Engineering,* vol. 144, pp. 135-141, 2017.

[172] Golovin, K., Boban, M., Mabry, J. M., and Tuteja, A., "Designing Self-Healing Superhydrophobic Surfaces with Exceptional Mechanical Durability," *ACS Applied Materials & Interfaces,* vol. 9, no. 12, pp. 11212-11223, 2017.

[173] Yilgor, I., Bilgin, S., Isik, M., and Yilgor, E., "Facile preparation of superhydrophobic polymer surfaces," *Polymer,* vol. 53, no. 6, pp. 1180-1188, 2012.

[174] Barshilia, H. C., Ananth, A., Gupta, N., and Anandan, C., "Superhydrophobic nanostructured Kapton® surfaces fabricated through Ar+ O2 plasma treatment: Effects of different environments on wetting behaviour," *Applied Surface Science,* vol. 268, pp. 464-471, 2013.

[175] Salapare, III, H. S., Guittard, F., Noblin, X., de Givenchy, E. T., Celestini, F., and Ramos, H. J., "Stability of the hydrophilic and superhydrophobic properties of oxygen plasma-treated poly (tetrafluoroethylene) surfaces," *Journal of Colloid and Interface Science,* vol. 396, pp. 287-292, 2013.

[176] Arianpour, F., Farzaneh, M., and Jafari, R., "Hydrophobic and ice-phobic properties of self-assembled monolayers (SAMs) coatings on AA6061," *Progress in Organic Coatings,* vol. 93, pp. 41-45, 2016.

[177] Menini, R. and Farzaneh, M., "Advanced icephobic coatings," *Journal of Adhesion Science and Technology,* vol. 25, no. 9, pp. 971-992, 2011.

[178] Choi, S. J., Suh, K. Y., and Lee, H. H., "A geometry controllable approach for the fabrication of biomimetic hierarchical structure and its superhydrophobicity with near-zero sliding angle," *Nanotechnology,* vol. 19, no. 27, p. 275305, 2008.

[179] Wang, Y., Xue, J., Wang, Q., Chen, Q., and Ding, J., "Verification of icephobic/anti-icing properties of a superhydrophobic surface," *ACS Applied Materials & Interfaces,* vol. 5, no. 8, pp. 3370-3381, 2013.

[180] Wu, L., Zhang, J., Li, B., Fan, L., Li, L., and Wang, A., "Facile preparation of super durable superhydrophobic materials," *Journal of Colloid and Interface Science,* vol. 432, pp. 31-42, 2014.

[181] Zhang, J., Li, J., and Han, Y., "Superhydrophobic PTFE surfaces by extension," *Macromolecular Rapid Communications,* vol. 25, no. 11, pp. 1105-1108, 2004.

[182] Nilsson, M. A., Daniello, R. J., and Rothstein, J. P., "A novel and inexpensive technique for creating superhydrophobic surfaces using Teflon and sandpaper," *Journal of Physics D: Applied Physics,* vol. 43, no. 4, p. 045301, 2010.

[183] Nosonovsky, M., and Rohatgi, P. K., *Biomimetics in materials science: self-healing, self-lubricating, and self-cleaning materials*. Springer Science & Business Media, 2011.

[184] Wenzel, R. N., "Resistance of solid surfaces to wetting by water," *Industrial & Engineering Chemistry,* vol. 28, no. 8, pp. 988-994, 1936.

[185] Cassie, A., and Baxter, S., "Wettability of porous surfaces," *Transactions of The Faraday Society,* vol. 40, pp. 546-551, 1944.

[186] Marmur, A., "Wetting on hydrophobic rough surfaces: to be heterogeneous or not to be?," *Langmuir,* vol. 19, no. 20, pp. 8343-8348, 2003.

[187] Guo, Z., and Yang, F., *Surfaces and Interfaces of Biomimetic Superhydrophobic Materials*. John Wiley & Sons, 2017.

[188] Lee, S. E., Lim, H. G., Lee, S. S., Choi, D. G., Lee, D., and Hong, S. U., "Embossed superhydrophobic polymer surfaces with topological variances," *Macromolecular Research,* vol. 21, no. 8, pp. 916-920, 2013.

[189] Hong, D., Ryu, I., Kwon, H., Lee, J. J., and Yim, S., "Preparation of superhydrophobic, long-neck vase-like polymer surfaces," *Physical Chemistry Chemical Physics,* vol. 15, no. 28, pp. 11862-11867, 2013.

[190] Xu, Q. F., Mondal, B., and Lyons, A. M., "Fabricating superhydrophobic polymer surfaces with excellent abrasion resistance by a simple lamination templating method," *ACS Applied Materials & Interfaces,* vol. 3, no. 9, pp. 3508-3514, 2011.

[191] Bormashenko, E., Grynyov, R., Chaniel, G., Taitelbaum, H., and Bormashenko, Y., "Robust technique allowing manufacturing superoleophobic surfaces," *Applied Surface Science,* vol. 270, pp. 98-103, 2013.

[192] Park, S. H., Lee, S., Moreira, D., Bandaru, P. R., Han, I., and Yun, D. J., "Bioinspired superhydrophobic surfaces, fabricated through simple and scalable roll-to-roll processing," *Scientific Reports,* vol. 5, p. 15430, 2015.

[193] Jafari, R., Momen, G., and Farzaneh, M., "Durability enhancement of icephobic fluoropolymer film," *Journal of Coatings Technology and Research,* vol. 13, no. 3, pp. 405-412, 2016.

[194] Yeong, Y. H., and Gupta, M. C., "Hot embossed micro-textured thin superhydrophobic Teflon FEP sheets for low ice adhesion," *Surface and Coatings Technology,* vol. 313, pp. 17-23, 2017.

[195] Huovinen, E., Hirvi, J., Suvanto, M., and Pakkanen, T. A., "Micro–micro hierarchy replacing micro–nano hierarchy: a precisely controlled way to produce wear-resistant superhydrophobic polymer surfaces," *Langmuir,* vol. 28, no. 41, pp. 14747-14755, 2012.

[196] Lee, S. E., Lee, K. W., Kim, J. H., Lee, K. C., Lee, S. S., and Hong, S. U., "Mass-producible superhydrophobic surfaces," *Chemical Communications,* vol. 47, no. 43, pp. 12005-12007, 2011.

[197] Cao, B. Y., et al., "Superhydrophobicity of self-organized surfaces of polymer nanowire arrays fabricated by a nano-injection moulding technique," *Journal of Thermal Science and Technology,* vol. 6, no. 2, pp. 204-209, 2011.

[198] Rasilainen, T., Suvanto, M., and Pakkanen, T. A., "Anisotropically microstructured and micro/nanostructured polypropylene surfaces," *Surface Science,* vol. 603, no. 14, pp. 2240-2247, 2009.

[199] Weng, C., Wang, F., Zhou, M., Yang, D., and Jiang, B., "Fabrication of hierarchical polymer surfaces with superhydrophobicity by injection molding from nature and function-oriented design," *Applied Surface Science,* vol. 436, pp. 224-233, 2018.

[200] Mielonen, K., Suvanto, M., and Pakkanen, T., "Curved hierarchically micro–micro structured polypropylene surfaces by injection molding," *Journal of Micromechanics and Microengineering,* vol. 27, no. 1, p. 015025, 2016.

[201] Hopmann, C., Recht, U., and Behmenburg, C., "Injection moulding of durable superhydrophobic surfaces made of LSR," *International Polymer Science and Technology,* vol. 41, no. 10, p. T1, 2014.

[202] Vazirinasab, E., Jafari, R., and Momen, G., "Application of superhydrophobic coatings as a corrosion barrier: A review," *Surface and Coatings Technology,* 2017.

[203] Huovinen, E., Takkunen, L., Korpela, T., Suvanto, M., Pakkanen, T. T., and Pakkanen, T. A., "Mechanically robust superhydrophobic polymer surfaces based on protective micropillars," *Langmuir,* vol. 30, no. 5, pp. 1435-1443, 2014.

[204] Yan, Z., Liang, X., Gao, Y., and Liu, Y., "Aging and self-healing properties of superhydrophobic silicone rubber," *IEEE Transactions on Dielectrics and Electrical Insulation,* vol. 23, no. 6, pp. 3531-3538, 2016.

[205] Yan, Z., Liang, X., Wu, C., Bao, W., Li, S., and Liu, Y., "Aging and recovery of superhydrophobic silicone rubber under electrical and non-electrical stresses," in *2015 IEEE Conference on Electrical Insulation and Dielectric Phenomena (CEIDP).*

Biographical Sketches

Khosrow Maghsoudi

Affiliation: Department of Applied Sciences, University of Quebec in Chicoutimi (UQAC), Chicoutimi, Quebec, Canada.

Education: BSc and MSc in Polymer Engineering, University of Tehran, Tehran, Iran. PhD Student, Department of Applied Science, University of Quebec at Chicoutimi (UQAC).

Business Address: 555, Boulevard de l'Université, Chicoutimi (Québec), Canada G7H 2B1.

Research and Professional Experience:

Teaching Assistant:

"Polymer Composites," Faculty of Engineering, University of Tehran.

"Mechanical and Physical Properties of Polymers," Faculty of Engineering, University of Tehran.

Research Assistant:

"Laboratory of Mechanical and Physical Properties of Polymers," University of Tehran.

"Laboratory of Mechanical Engineering and Composite Materials," University of Quebec at Chicoutimi (UQAC).

Honors:

Ranked 17th in "National Universities Entrance Exam – Graduate" among more than 4000 participants.

Awarded full scholarship for the graduate and undergraduate program, College of Engineering, University of Tehran.

Ranked 2nd in all M.Sc. students of polymer engineering, College of Engineering, University of Tehran.

Awarded for successfully completing the "MATLAB," "Transmission Electron Microscopy (TEM)," "Design of Experiment (DOE)" courses.

Awarded Bourse d'excellence – HYDRO-QUEBEC de la Fondation de l'Universite du Quebec a Chicoutimi 2016-2017.

Awarded Bourses du Fonds de développement de l'UQAC, Excellence cycles supérieurs, 2016-2017.

Awarded Bourse d'excellence – HYDRO-QUEBEC de la Fondation de l'Universite du Quebec a Chicoutimi 2017-2018.

Awarded Bourse d'excellence – Masoud Farzaneh de la Fondation de l'Universite du Quebec a Chicoutimi 2017-2018.

Publications from the Last 3 Years:

Journal Articles

"Absorption of Heavy Metals Using Resorcinol Formaldehyde Aerogel Modified with Amine Groups," Siamak Motahari, Mahmoud Nodeh, Khosrow Maghsoudi, *Desalination and Water Treatment*, Taylor and Francis, 2015.

"Mechanical, Thermal and Hydrophobic Properties of Silica Aerogel-Epoxy composites," Khosrow Maghsoudi, Siamak Motahari, *Applied Polymer Science*, Wiley, 2017.

"Micro-nanostructured polymer surfaces using injection molding-A review," K. Maghsoudi, R. Jafari, G. Momen, M. Farzaneh, *Materials Today Communications*, 2017.

Conference Papers

"Mechanical properties of epoxy/fumed silica nanocomposites," Khosrow Maghsoudi, Siamak Motahari, Accepted for Oral Presentation in *Sustainable Materials Science and Technology*, An International Conference, Paris, 15-17 Jul. 2015.

"Improvement of Compression Strength of Silica Aerogel Using Poly Ethylene Glycol," Siamak Motahari, Khosrow Maghsoudi, Navid Eskandari, Mojtaba Kazemi, Accepted for Oral Presentation in *The 15th Iranian National Congress of Chemical Engineering*, Tehran, 17-19 Feb. 2015.

"Micro-nanostructured silicone rubber using direct replication," Khosrow Maghsoudi, Reza Jafari, Gelareh Momen, Masoud Farzaneh, *1st Annual QCAM Symposium,* University of Sherbrooke, Canada, October 2017.

"Micro-Nanostructured Silicone Rubber Surfaces Using Compression Molding," Khosrow Maghsoudi, Reza Jafari, Gelareh Momen, Masoud Farzaneh, Tony Carreira, *THERMEC'2018 International Conference on Processing & Manufacturing of Advanced Materials*, July 8-13, 2018, Paris, France.

Reza Jafari, PhD

Dr. Reza Jafari is a professor-researcher in the department of Applied Sciences at Université du Quebec à Chicoutimi. He has a BSc and MSc in chemical engineering from Isfahan University of Technology and Sharif University of Technology in Iran. He received his PhD in Process Engineering and High Technology in 2006 from Université Pierre et Marie Curie (Paris 6), France. His research interests are surface treatment, composite materials and nanostructured surfaces. He is involved in a several projects on the development of multifunctional and smart coatings.

Gelareh Momen, PhD

Gelareh Momen has a BSc and MSc in chemical engineering from Sharif University and Polytechnic of Tehran in Iran, and PhD in process engineering form University of Paris XIII. Her research focuses on the development of nanostructured superhydrophobic/icephobic/anticorrosive surfaces by using different processes including: plasma technology, spray coatings, self-assembled coatings, anodization, and nanocomposites.

She is currently professor in University of Quebec in Chicoutimi. Prior to joining UQAC, she worked as R&D engineer at Bombardier Aerospace on interior coatings for aircraft and set up and equipped the first interior GCC experimental laboratory. She was also member of Working Group B2-44, Coatings for protecting overhead power network equipment in winter conditions Principles of Overhead Line Design. Her research interests include nanodielectric coating, ice and pollution adhesion on power network, polymeric insulator and self-cleaning coatings. She is the responsible of LARGIS laboratory (Laboratoire revêtement glaciophobe et ingénierie de surface).

Masoud Farzaneh

Masoud Farzaneh (M'83-SM'91-F'07) has done pioneering research in the field of power engineering, including the impact of cold climate on overhead power networks. He has authored or co-authored more than 700 technical papers, and 17 books or book chapters in power engineering. As mentor, Professor Farzaneh has trained about 130 postgraduate students and postdoctoral fellows to date. He was President of IEEE DEIS for 2013, member of the Editorial Board of IEEE Transactions on Dielectrics and Electrical Insulation, as well as Convenor of CIGRE WG B2 29 and B2.44

respectively on de-icing techniques and coatings for protection of overhead lines during winter conditions and member of the Executive Committee of CIGRE Canada. He is presently Editor-in-Chief of the IET High Voltage journal and Convenor of CIGRE WG B2 69 on coatings for power networks. Professor Farzaneh is Life Fellow of IEEE, Fellow of The Institution of Engineering and Technology (IET), Fellow of The Engineering Institute of Canada (EIC) and Fellow of The Canadian Academy of Engineering (CAE). His contributions and achievements in research and teaching have been recognized by many prestigious prizes and awards at national and international levels.

In: Polymer Surfaces
Editor: Valente Erasmo Silva
ISBN: 978-1-53614-309-6

Chapter 2

FABRICATION OF MULTILAYERED POLYELECTROLYTE COATINGS ON SACRIFICIAL CORES: TOWARDS SMART AND SIZE-TUNABLE DRUG-DELIVERY CAPSULES

***Sabrina Belbekhouche*[*], *Julie Oniszczuk* and *Benjamin Carbonnier*[†]**
Université Paris Est, ICMPE (UMR7182), CNRS, UPEC,
Thiais, France

ABSTRACT

Materials with high surface-to-volume ratio have opened new avenues in materials design for a plethora of scientific and technological areas, including energy storage and conversion, catalysis, sensing, biology and medicine. Beyond morphological aspects, the properties of high surface area materials such as polymeric nanoparticles can be tuned

[*]Corresponding author. Assistant Professor, Eastern Paris University, Thiais, France. E-mail: belbekhouche@icmpe.cnrs.fr; Tel: +331 4978 1149.
[†] Assistant Professor, Eastern Paris University, Thiais, France. Email: carbonnier@icmpe.cnrs.fr; +331 4978 1114.

through surface engineering strategies, and various methods have been reported in literature so far to design polymeric nanoparticles with well-defined topology, size, chemical composition, surface properties and related biodistribution pattern and bioavailability. Beyond the widely studied core-shell nanoparticles, nanocapsules, i.e., hollow nanoparticles, have emerged as a viable and versatile platform for drug delivery applications.

This contribution will focus on biocompatible and biodegradable hollow nanoparticles prepared *via* the layer-by-layer deposition of polymer thin films on sacrificial templates. The review will encompass all aspects of nanocapsules from the preparation, through the characterization and to the applications as drug carriers. In a first section, the growth mechanism (linear vs. exponential) of LbL films on particulate templates will be discussed and rationalized based on the nature (polyelectrolyte) and characteristics (molar mass, charge state) of the polymers. The formation of nanocapsules will be presented as a function of the nature of the particulate template (metal, metal oxide, latex) and its dissolution process. Main characteristics of nanocapsules will be illustrated with respect to the experimental characterization methods. Judiciously selected examples from literature will be highlighted to demonstrate major advantages and benefits of using nanocapsules as drug delivery systems, notably in terms of drug loading and targeting ability. The last section will discuss future prospects about biodegradable and biocompatible nanocapsules in disease treatment.

INTRODUCTION

Implementation of nanotechnology to medicine has completely reshaped our idea about therapy and diagnosis. Classical therapeutic approaches suffer from several side effects and poor efficacy. Progress and promise of nanomedicine, or more precisely in drug delivery systems (DDS), open new avenues to address therapy issues for cancer and other diseases treatment. Research at the intersection of different scientific and technological areas, such as biotechnology, medicine, engineering, nanotechnology, and material science is then required to allow the development of new and efficient approaches using tailor-made vectors [1]. For instance, conventional anticancer therapies are often inefficient due to poor efficacy and induce severe side effects. This can mainly be correlated to a lack of selectivity of therapeutic agents for the diseased tissues. Thereby, the administered drug interacts indiscriminately with the rest of the body affecting both healthy and diseased tissues yielding a decrease of the effective therapeutic dose (Figure 1). The development of stable nanosized vehicles acting as reservoirs is then crucial to maximize the pharmacokinetics and biodistribution of therapeutic agents.

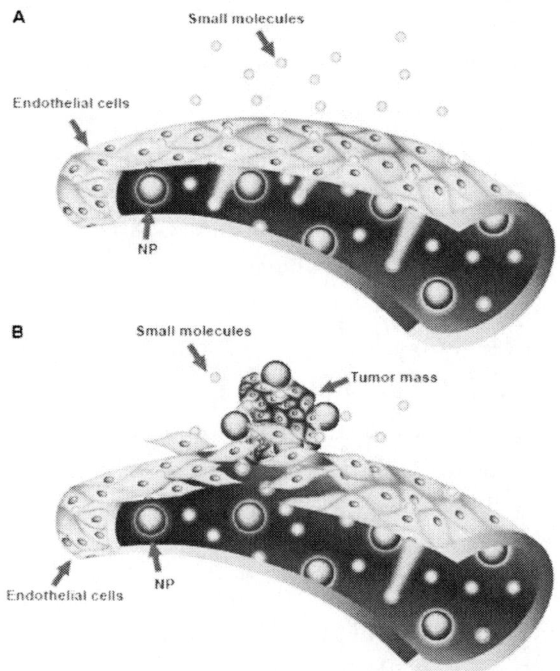

Figure 1. Targeting of nanomedicines by the enhanced permeability and retention (EPR) effect. Notes: Differences between normal (**A**) and tumor (**B**) vessels are depicted. Tumor vessels contain large fenestrations between the endothelial cells: this structural characteristic allows the nanoparticles (NPs) to reach the matrix and the tumor cells by the EPR effect. Conversely, normal tissue contains tightly joined endothelial cells: this prevents the diffusion of NPs outside the blood vessels [2].

On the contrary to free drugs which may diffuse non-specifically, drug-loaded nanocarriers with a size of about 100 nm [3] can target diseased tissues/cells. Then, size matters! Indeed, to fully gained advantage of the drug delivery effect, nanocarriers must meet specifications, in terms of size, but also biocompatibility, biodegradability, hydrophilicity…, to prevent unwanted phenomenon such as phagocytosis by macrophages or endocytic uptake of Kupfer cells which leads to the removal of the drug-loaded nanoparticles from the blood stream. Designing nanocarriers with the appropriate size is one of the major prerequisites to treat efficiently a disease as it has proved to reduce the recognition of nanoparticles by the immune system.

Among the different strategies for elaborating nanovectors, the design of nanometer-sized hollow polymeric spheres (or nanocapsules) is a new expanding area of physico-chemical research. This field has received increasing attention as it enables to access to a large range of capsules, with tuneable wall thickness, diameters, and permeability [4, 5]. Of utmost relevance to the biomedical field as in drug delivery/targeting, nanocapsules have the ability to encapsulate large quantities of guest molecules [6, 7]. Different methods have been reported to date to construct hollow nanoparticles (nanocapsules), in this chapter, we will only pay attention to the layer-by-layer (LbL) deposition of polyelectrolytes on a sacrificial template [8] also called "template method" [9-11]. Both preparation, physicochemical characterization, and applications of the said multilayered nanocapsules as drug carriers are considered. This preparation method allows not only creating hollow structured particles, but also the control of the particle shape and size, the shell permeability and thickness and the surface functionality.

The large surface area of the multilayered shell combined to the large void space in the core of the hollow particles provide unique opportunities to encapsulate bio(macro)molecules such as drugs, DNA or proteins while the LbL construction method enables the incorporation of specific building bricks for targeted release. Other key parameters have also to be considered and one may mention biocompatibility and biodegradability, stimuli-responsive behavior, drug release mechanism and biodistribution. Biopolymers such as proteins, carbohydrates and other biosynthetic polymers, are produced by living organisms, including bacterias, animals and plants. Their biodegradability and biocompatibility ensure safer applications in tissue engineering and drug delivery for instances [12, 13]. Hence, hollow nanoparticles designed from biopolymers may combine typical properties of their counterparts prepared from synthetic polymers, in terms notably of mechanical stability, and benefits due to the bioinspired nature as being intrinsically biodegradable, abundant in nature at reasonable cost, and renewable... Furthermore, biopolymers possess a high content of functional groups including amino, hydroxyl, and carboxylic acid groups for further chemical functionalization, as to bring stimuli-responsiveness.

The remote control of drug release can be achieved using thermo- [14], photo-[15], magnetic-[16] or ultrasound-sensitive nanovectors [17]. Moreover, it is also judicious to take advantage of specific micro-environmental variation associated with neoplastic diseases to stimulate spontaneously tailored release profiles. The most explored endogenous *stimuli* include pH [18], the altered redox potential due to higher glutathione concentration (thiol/disulfide chemistry) [19] and the up-regulation of certain enzymes [20].

In this chapter, the recent elaboration methods of hollow nanoparticles *via* the deposition multilayered polyelectrolyte coatings on sacrificial cores are summarized. More precisely, the first section will present the general features about LbL assembly methods. The growth mechanism and types of templates will also be discussed. The next section will highlight the major trigger chemistry used to provide controlled release. In the last part, a limited number of selected applications toward drug delivery will be presented.

I. LBL ASSEMBLY THROUGH ELECTROSTATIC INTERACTION FOR THE DESIGN OF HOLLOW NANOPARTICLES

I.1. General Overview

Since the pioneering work of Decher and co-workers [21-23], numbers of publications based on the implementation of the multilayered polyelectrolytes concept have been reported [24-26]. Indeed, LbL assembly driven by electrostatic interactions enables to elaborate multilayer polyelectrolyte films with well-controlled structure, thickness and composition, simply via the alternate deposition of oppositely charged polymers onto a charged substrate (Figure 2) [22]. The method can be achieved in aqueous solutions and is well-suited for the elaboration of ultrathin and homogeneous films providing efficient solutions for several applications including biomedical ones.

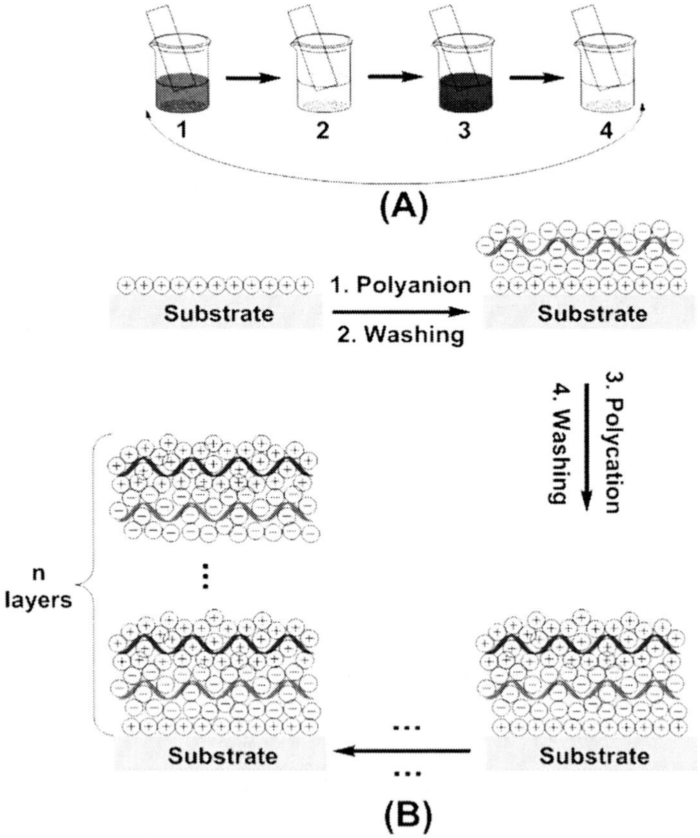

Figure 2. Schematic representation of the buildup of multilayer thin films via electrostatic interactions: (A) immersion of the charged substrate in the corresponding polyionic solution, with alternating washing steps; (B) schematic representation of the construction of the polyionic architecture depicting film deposition starting, for example, with a positively charged substrate. The successive immersions (n times) of the substrate in the polyion solutions allow the buildup of a multilayer film at the molecular level (n layers) [27].

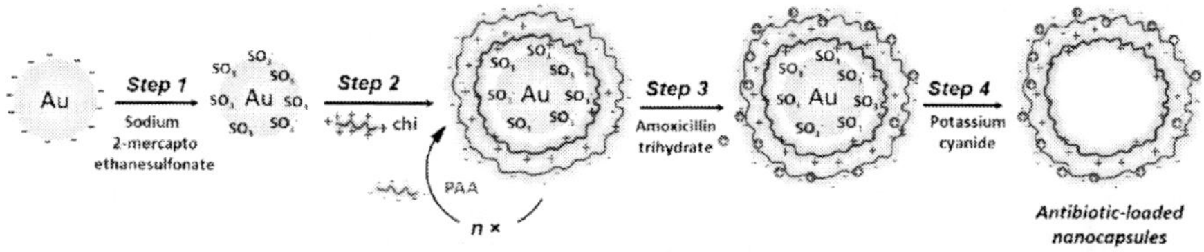

Figure 3. Schematic illustration of the elaboration pathway for the preparation of drug-loaded hollow nanocapsules involving; step 1: preparation of sulfonate-decorated gold nanoparticles, step 2: assembly of chitosan/poly(acrylic acid) multi-layered films, step 3: antibiotic loading and step 4: hydrolysis of the gold core [30]. For clarity purpose the scheme is not to scale.

The LbL technique is one of the most promising methods to elaborate multifunctional nanocapsules. In this case, the successive deposition of polymer layers is achieved on the surface of a given template enabling structural control at the macromolecular scale [9, 11, 28, 29]. Finally, the selective removal of the templates leads to the formation of hollow nanoparticles. The main driving force for the assembly build-up remains the electrostatic attraction between polyelectrolyte layers of opposite charge (polyanion and polycation, as shown in Figure 2), which makes the deposition procedure very simple and straightforward to modify spherical surfaces [22].

As illustrated in Figure 3, the LBL process for designing nanocapsules typically implies four steps starting from the template. One of the most common challenge is the limited compatibility between the template surface and the polyelectrolyte shell. This may result in the self-aggregation of the polymers yielding formation of complex and thus, the unsuccessful shell formation. Step 1 relies on a surface treatment of the substrate's surface to ensure the robust anchoring of the polyelectrolyte. It may involve the adsorption or the grafting of molecules with low molecular weight or macromolecules such as polyethyleneimine, mercaptoethane sulfonate, (meth)acrylate or vinyl units. In case of templates with high charge density, step 1 may be omitted. Step 2 is the repetition of adsorption cycles of a polyelectrolyte pair, namely a combination of polycation and polyanion. Steps 3 and 4 involve the loading of the releasable drug and dissolution of the template core, respectively. Step 3 and 4 can be achieved in reversed order.

I.2. LbL Growth Mechanism: Linear versus Exponential Growth

Polyelectrolytes multilayer films can grow linearly or exponentially, the latter providing thicker films. A transition regime between both types of growth has been reported according to the strength of the polyelectrolyte intermolecular interactions. The linear growth regime has been widely investigated in the literature compared to the exponential one as this latter is a more recent approach. The linear growth is to be correlated to both linear increase of the mass and film thickness with the increase of the number of deposited polyelectrolyte layer [31]. This could be correlated to the non-diffusion of the polyelectrolytes constituting the multilayer film. A linear growth is typically seen for films composed by strong polyelectrolyte pairs such as poly(allylamine hydrochloride) and poly(styrene sulfonate) [32, 33].

Generally, polyelectrolyte pairs based on at least one biopolymer (polysaccharide or polypeptide) have been evidenced to exhibit exponential growth. Examples include poly(L-lysine)/alginate (PLL/ALG), PLL/hyaluronan (PLL/HA), PLL/poly(acrylic acid) (PLL/PAA), PLL/poly(L-glutamic acid) (PLL/PGA), chitosan/Alginate (CHI/ALG), polyethylenimine/PAA (PEI/PAA) or PGA/PAH, chitosan/HA (CHI/HA) [34-42]. In these cases, mass and thickness grow with the number of the adsorbed polyelectrolyte faster than in linear systems.

In the literature, the diffusion of the polycationic chains "in-and-out" of the whole film during the multilayer buildup has been highlighted as the main reason for the exponential growth [43], and some simple models have been proposed. For instance, Salomäki and Kankare [44] proposed an acoustic-based theoretical model which unveiled that the growth mode of the well-known CHT/HA and PLL/HA multilayers is mainly exponential. Moreover, Hoda and Larson [45] suggested taking account of the diffusion of polycations "in-and-out" of the film as well as film dissolution and the existence of an energetic barrier at the film surface. Other physicochemical parameters e.g., adsorption time, ionic strength and nature of the electrolyte species, temperature, pH, charge density, average molar mass, and conformation of the polyelectrolytes, concentration of the polyelectrolyte solution have been demonstrated to also impact final characteristics of polyelectrolyte multilayer assemblies such as thickness, roughness, permeability, wettability, swelling behavior to name but a few… [31, 40, 46-49]. For a more in-depth discussion on this section, the readers is kindly invited to refer to an excellent recent review by Borges and Mano [27].

I.3. Templates Used for the Elaboration of Capsules

In this section, we only discuss about hollow particles (nanocapsules) obtained though surfaces coating of colloidal templates with thin layers of polymeric materials, followed by selective removal of the templates. As illustrated in Figure 3, this process may rely on multiple steps. The most common challenge is the incompatibility between the template surface and polyelectrolyte shell. This may result in the self-aggregation of the polyelectrolytes yielding complex formation, and thus the unsuccessful formation of shell. Nanocapsules prepared through LbL strategy exhibit highly controlled dimensional features as evidenced by advanced morphological characterization. This control is directly correlated to the template used, as well as to the nature of the polyelectrolyte chains that can be interdispersed with different molecules to produce a capsule's wall with desired functionality and composition. Moreover, the direct coating of particulate templates via the LbL approach enables achieving uniform coating with tunable thickness. Indeed, since the pioneering work by Caruso *et al.* who were the first to extend the LbL technique to elaborate hollow inorganic silica and hybrid capsules [8], this simple and straightforward approach has been applied to a variety of materials, including organic or inorganic templates as discussed here below.

I.3.1. Gold Nanoparticles as Templates

Gold nanoparticles combine very interesting properties, they are biocompatible, commercially available in a wide range of size including diameter as low as 5 nm, provide an excellent contrast in computerized tomography and electron microscopy, gold quantum dots exhibit intrinsic fluorescence. The deposition of successive polyelectrolyte multilayers onto gold nanoparticles [6, 50] can be easily monitored by exploiting the sensitivity of the gold nanoparticle's surface plasmon band due to the adsorbed species [51]. The mechanism by which the colloids are stabilized is based on electrostatic interactions [52], and the main issue consists in avoiding the polyelectrolyte bridging between several particles, resulting in a flocculation phenomenon (Figure 4).

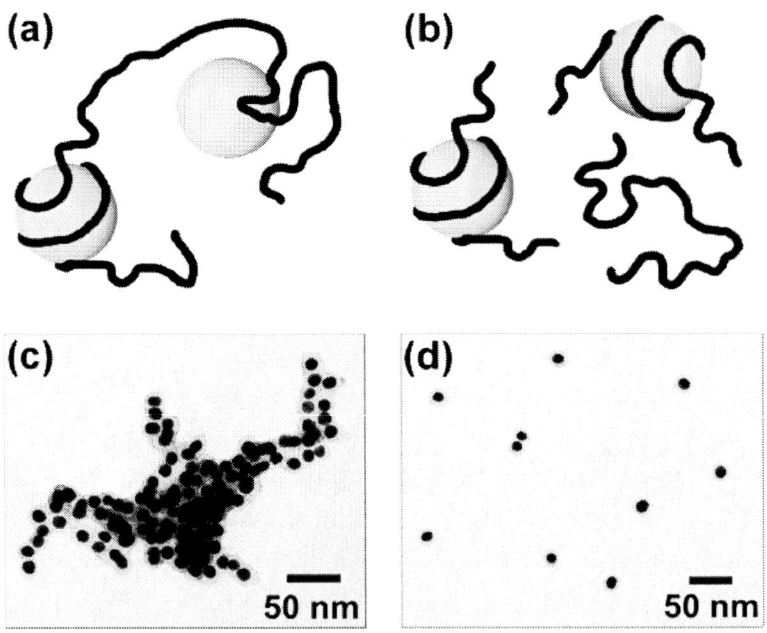

Figure 4. Oversimplified representations of particles being aggregated by bridging flocculation (a) and being individually wrapped by polymer chains (b); lower part shows electron micrographs of aggregated particles (c) and of particles predominantly unsheathed individually (d) [52].

Dissolving the sacrificial Au^0 core allows to obtain nanocapsules through the formation of water-soluble gold complex. Potassium cyanide is typically used to enable gold hydrolysis. The cyanide-assisted etching leads to the formation of a water-soluble salt of potassium gold cyanide complex (equation 1).

$$4\ Au + 8\ KCN + 2\ H_2O + O_2 \rightarrow 4\ KOH + 4\ KAu(CN)_2 \qquad \text{Eq. 1}$$

The $KAu(CN)_2$ complex can easily be eliminated through diffusion across the polyelectrolyte membrane [53, 54]. Such process of core dissolution could easily be monitored by spectrophotometric monitoring of the disappearance of the plasmon absorbance band. Transmission electron microscopy (TEM), atomic force microscopy (AFM) and electrophoresis were also employed to show the deposition of the polyelectrolyte layers onto the gold matrix.

Detailed investigation demonstrated the exponential trend for the diffusion through the shell that could be correlated to diffusion from a sphere as previously reported in the literature [6, 55]. The diffusion coefficient was extracted by using the following equation (equation 2) [6]:

$$1 - (C_t - C_0)/(C_s - C_0) = b\ \exp(-At) \text{ with } A = D\pi^2/L^2 \qquad \text{Eq. 2}$$

where C_t is the concentration of diffusing species at time t, C_s is the surface concentration, C_0 is the initial concentration of diffusing species at the film/particle interface, and b is a constant dependent on the initial boundary conditions implied in solving Fick's laws. C_0 may be fixed to 0 and the left hand side of equation 3 may be correlated to the gold Plasmon intensity.

$$(abs_t/abs_0) = b\ \exp(-At) \qquad \text{Eq. 3}$$

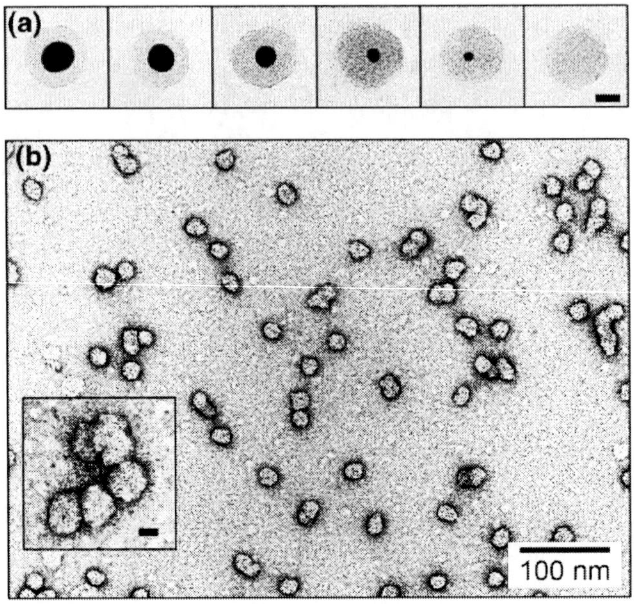

Figure 5. (a) Unstained transmission electron micrographs of individual Au-core/polyelectrolyte shell nanoparticles at different stages of the dissolution of the Au core. The polyelectrolyte shell had a thickness of 20 layers. The bar in the right image corresponds to 10 nm. (b) Overview transmission electron micrograph of "empty nanospheres" after the dissolution of the gold core and staining with uranyl acetate. In this sample the polyelectrolyte shell was composed of eight layers. The inset shows an aggregate of 4 or 5 particles at higher magnification (the bar corresponds to 10 nm) [56].

Typically, it was shown that the lower the size of the gold nanoparticle, i.e., the core template, the faster the gold etching process.

Gittins and Caruso [53, 55] investigated the elaboration of nanocapsules based on sodium poly(styrene sulfonate) (PSS) and poly(diallyldimethylammonium chloride) (PDADMAC) by using gold nanoparticles with different diameters (11 ± 2 nm to 54 ± 13 nm). The polyelectrolyte pair was deposited on the template and the polymer excess was removed by centrifugation process. Schneider and Decher [56] have shown how gold colloids, in concentrated suspension, can be coated with up to 20 layers of the polycation poly(allylamine hydrochloride) (PAH) and the polyanion poly(styrene sulfonate) (PSS) and subsequently chemically degradaded (Figure 5). The authors monitored changes in the plasmon band of the gold colloids in the aim to prove the adsorption of the polyelectrolyte layer on the colloidal surface (red shift of the absorption band from 0.5 nm to about 10 nm at the saturation limit) and to discriminate between interparticle bridging and aggregation (red shift of typically more than 80 nm to about 150 nm).

Belbekhouche and co-workers reported on the preparation of ultra-low sized (20 and 60 nm) biodegradable polymeric capsules based on the chitosan/poly(acrylic acid) and exhibiting antibacterial activity [30]. The authors judiciously gained advantage of the electrostatic interactions to construct the multilayer films but also to load a small molecule containing amine group, namely amoxicillin. Crosslinked capsules were also elaborated through intermolecular reaction between the chitosan chains using glutaraldehyde as crosslinker. The authors also evidenced the possibility to control the amount of loaded amoxicillin through controlling the size of the gold template, the thickness and crosslinking density of the multilayer films.

I.3.2. Metal Oxide Nanoparticles as Templates

Silica particles are commercially available with a wide range of sizes (approximately 0.1 to 10 μm) and surface functionalities. The preparation of capsules based on silica particles requires to dissolve the core template, after the self-assembly, using hydrofluoric acid. This approach has been initially developed by Caruso et al. and successfully employed to fabricate capsules loaded with therapeutics, e.g., functional DNA or protein [57]. Two processes were involved by Yu et al., i.e., electrostatic interactions and chemical complexation processes, to elaborate a novel iron-heparin complexed hollow capsule. These capsules were fabricated by alternating deposition of ferric ions (Fe^{3+}) and heparin (polyanion) onto a surface of polystyrene latex particles with two different sizes (488 nm and 10.55 μm), followed by the cores removal by tetrahydrofuran [58].

I.3.3. Calcium Carbonate Nanoparticles as Templates

The use of calcium carbonate as template is an appealing alternative. $CaCO_3$ can be easily synthesized, it is non-toxic and can be dissolved under slightly acidic conditions or by complexation with ethylenediamine tetraacetic acid (EDTA) [59]. Calcium carbonate particles have a porous structure that can be employed to encapsulate chemicals with a high loading capacity. Nagaraja et al. [60]. reported the elaboration of nanosized porous calcium carbonate which remains stable over time in solution. The addition of poly(vinylsulfonic acid) (PVSA) during the synthesis process allows to lower the nucleation rate and then stabilize the surface to prevent further growth or aggregation into particles. Biswas et al. [61] judiciously fabricated oxygen-sensitive polymer nanocapsules based on PVSA-stabilized calcium carbonate nanoparticle as templates (Figure 6). The authors examined the conditions for LbL deposition of (poly(sodium 4-styrenesulfonate/poly(diallyldimethylammonium)) (PDADMAC/PSS) on $CaCO_3$ colloidal templates followed by EDTA treatment for the elaboration of hollow nanoparticles.

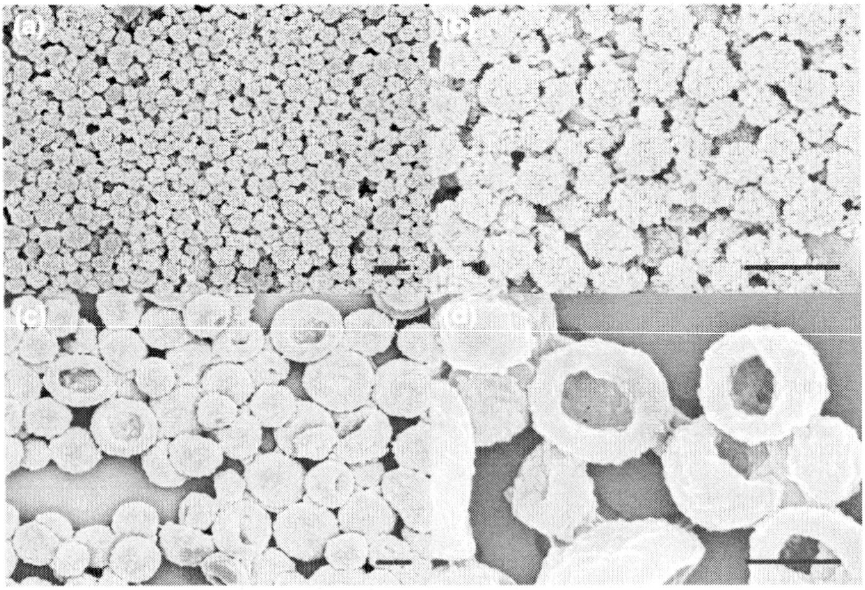

Figure 6. SEM images of sputter-coated CCNPs at (a) 20 000× magnification (b) 50 000× magnification and NCs at (c) 20 000× magnification and (d) 50 000× magnification. Scale bars correspond to 500 nm [61].

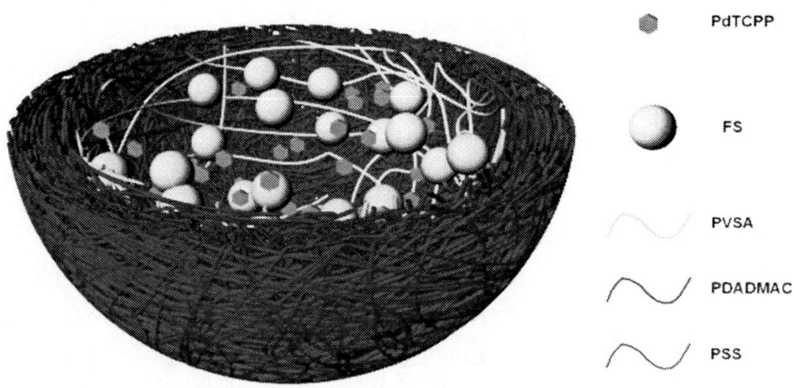

Figure 7. Representation Showing the Cross-Sectional View of a Nanocapsule Containing FS and PdTCPP [61]. FS and PdTCPP meant for carboxylate modified FluoSpheres and Pd-meso-tetra(4-carboxyphenyl) porphine, respectively.

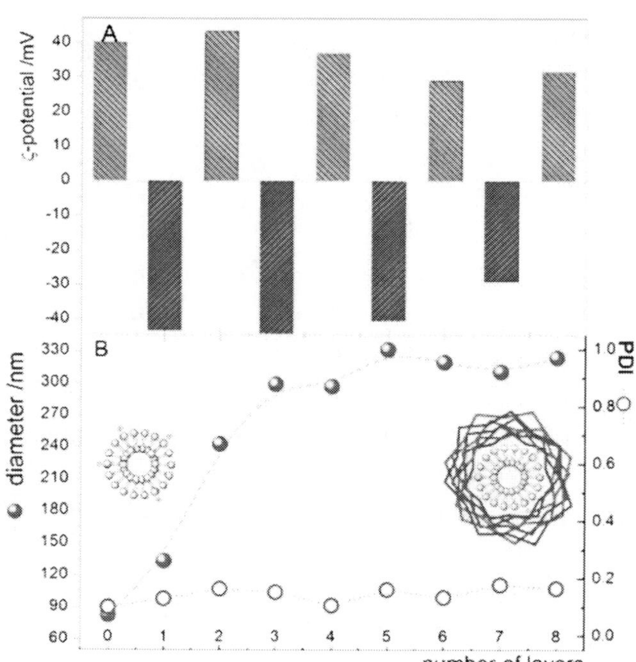

Figure 8. z-Potential values alternate with the adsorption of polyelectrolyte layers (panel A); average particle size (full dots) and polydispersity index (PDI) (empty dots) on the adsorption of polyelectrolyte layers onto the liposomes (panel B). The first values in both panels refer to the bare liposome [64].

Interestingly, Biswas et al. [61] further investigated the potential to encapsulate functional materials such as reporter dyes and nanoparticles (i.e., by employing the template-based method in conjunction with the LbL approach (see Figure 7).

I.3.4. Vesicle/Liposome as Templates

The self-assembly of amphiphilic molecules allows the formation of micelles or vesicles with different structures with the proviso that the concentration in aqueous solution exceeds critical values [62]. F. Cuomo and coworkers [63] fabricated hollow nanocapsules starting from didodecyldimethylammonium bromide (DDAB) vesicles. The authors amended their approach to obtain smaller hollow structures of 300 nm in diameter by implementing the phosphatidylcholine/DDAB liposome system as template which was coated by chitosan and alginate, as polyelectrolytes pair (Figure 8) [64]. The core– shell and hollow capsules were studied by means of Dynamic Light Scattering (DLS), z-potential and scanning electron microscopy (SEM).

I.3.5. Liquid Template as Templates

When two immiscible liquids are mixed together through mechanical agitation, liquid droplets are generated forming an emulsion that is a dispersed phase in a continuous phase. Liquid-liquid interfaces possess unique properties, e.g., high interfacial tension and fluidity. They could be used as an ideal template for the preparation of polyelectrolyte multilayer nanocapsules for drug delivery. The so-designed nanosized systems combine tunable shell properties and a liquid oily core able to contain high amounts of hydrophobic payloads. From a thermodynamic point of view, the self-assembly of polyelectrolytes at liquid/oil interface may be challenging due to the high reactivity and intrinsic instability [65]. This can be rationalized by the immiscible nature of oil and water phases, and the high interfacial tension. Similar effect can be observed even in presence of an emulsifier acting as a stabilizing agent. However, this drawback can easily be overcome by reducing surface tension through interaction with adjacent (macro)molecules or colloidal particles [66-69]. This is particularly relevant when multiple polymer depositions are considered, as in the Layer-by-Layer (LbL) assembly process [70-74]. In contrast to the use of solid template, the main advantage of depositing polyelectrolytes on a liquid template is that removal of the liquid drop is not required, avoiding tedious and time consuming steps.

II. Major Trigger Chemistries for Controlled Drug Release

Drugs' encapsulation is usually envisioned to achieve controlled release of the active molecules. One particularly efficient strategy relies on the use of triggers to induce release, i.e., upon a specific stimulus, at the desired location and time of action. It is now well-admitted that the use of smart delivery systems that undergo morphological changes in response to varied physiological environments can enhance the efficiency of the release in terms notably of release kinetics, dose and location. Herein, we will briefly discuss the most applied trigger based on the use of heat, light, redox effects, and pH to achieve active release. Release has also been successfully triggered by other mechanisms such as mechanical shear [75], ultrasound activation [76], enzymes [77].

II.1. Thermally-Triggered Chemistry

Thermally-responsive drug delivery systems are among the most studied mainly because tumor tissues exhibit abnormal temperature gradients as compared with normal ones and they are highly sensitive to temperature increases [78]. As proved by Issels R.D. the combined use of hyperthermia and chemotherapy is highly beneficial for cancer treatment, and effective hyperthermia-assisted killing of tumors can occur in the range 40-43°C [79]. Thermally-induced release is generally accompanied with phase changes, structural disruptions, diffusion enhancement, or a combination of these effects. Since the pioneering studies on the thermoresponsive behavior of acrylamide-containing polymers, poly(N-isopropylacrylamide) (PNIPAAM) has been widely investigated [80, 81]. Indeed PNIPAAM undergoes changes in hydrophilicity upon the temperature-dependent change conformation from coil-to-globule. Such a transition enables to controlling the solubility, and thus permeability of the PNIPAAM delivery system. Reversible thermal transition of Pluronic F127, that is made of poly(ethylene oxide)–poly-(propylene oxide)–poly(ethylene oxide) (PEO–PPO–PEO) and its derivatives have also been largely applied [82, 83].

II.2. Light-Triggered Chemistry

Photochemical approaches as compared to their pH, redox and thermal counterparts offer unique opportunities to achieve both spatial, temporal and remote control over the triggering process. Spatio-temporal and remote control of bioactivity such a drug release is a key step towards clinical and scientific breakthroughs. Different photochemical mechanisms have been identified so far as enabling the fine control of photoresponsive nanocarriers, behavior and one may cite surface plasmon absorption and related photothermal effects (Photoexcitation of metal nanostructures via strong absorption in the NIR region inducing he formation of a heated electron gas, i.e., localized heating), photo-induced isomerization (conformational change around a rotationally restricted bond), photo-driven oxidation (formation of singlet oxygen as a strong oxidizing agent upon irradiation of a sensitizer molecule), photo-controlled changes in hydrophobic/hydrophilic balance (disintegration of micellar systems through conversion of amphiphilic molecules to more hydrophilic ones), photo-conducted fragmentation of polymer chains (degradation of polymer backbones into smaller moieties providing also the disassembly of the nanocarrier), and photo-assisted de-crosslinking (breaking of crosslinking units). The literature is very rich and the reader is referred to the very exhaustive review [84].

Besides the physical stimuli discussed here above, chemical stimuli including pH, ionic strength, solvent, chemical and electrochemical response have also been investigated. Here below are presented the main features of trigger chemistries based on changes in pH and redox effects.

II.3. pH-Triggered Chemistry

The implementation of pH as a trigger suggests the possibility to induce chemical reaction and/or change if materials' properties via a change in the concentration of hydronium (H_3O^+) ions. Variations of pH are observed in nature, the most obvious being the change from acidic environment in the stomach to more neutral pH in the intestines. Variations from the physiological pH value of 7.4 to lower pH values is also for instance observed in extracellular environment of tumor tissue (pH 6.6–7.4) and endosome (pH = 5).

Both pH pH-catalyzed cleavage of covalent bonds as well as changes in hydrophilicity/hydrophobicity upon protonation/deprotonation equilibrium can be used to design pH-triggered materials. Previously reported examples include acetals, esters, boronic acids, imines, hydrazones and esters functional groups. To ensure release phenomenon, the cleavage of acid/base-sensitive bonds must yield change in permeability which is often achieve via depolymerization, structural fragmentation, or polarity change. In contrast, polyelectrolyte-based encapsulants are sensitive to pH variations and the corresponding changes in ionization state through protonation or deprotonation, modifies the electrostatic interaction forces and the overall hydrophilicity. Combination of both effect has been proposed by Almutairi and co-workers who designed nanoparticles from a copolymer of ketal and aminoester monomers [85]. The nanoparticles exhibited a burst release profile at lower pH while protonation induced a swelling effect providing an acceleration of the ketal cleavage throughout the nanoparticles.

II.4. Redox-Triggered Chemistry

Although cysteine is one of the least abundant amino acids, its presence is frequently associated within biological functional (regulatory, catalytic or binding) in proteins. This feature can be rationalized by the unique chemistry of the thiol (or thiolate) of cysteine exhibiting a pronounced nucleophilic character, high affinity for metal, and/or ability to form disulfide bonds. The latter characteristics if of particular interest regarding smart materials as thiol–disulfide (SH-S–S) chemistry is both exchangeable and reversible [86] A series of low molecular weight thiols has been identified as key actors in redox-mediated processes in cells, and the most ubiquitous is glutathione (GSH) a tripeptide made of Glu, Cys and Gly. Of utmost interest, cells exhibiting a high proliferation rate, typically cancer cells, possess a high level of glutathione especially in the nucleus [87]. In the cellular environment, GSH acts as a reducing agent, and may disassemble nanocarriers through the cleavage of the S-S linkage inducing the release of drugs [88].

III. Applications of Smart Nanocapsules to Drug Delivery

Hollow nanoparticles possess intrinsic characteristics such as high surface-to-volume ratio, low density refractive indices and thermal expansion coefficients. Moreover, through the judicious choice of the building bricks it is possible to provide additional properties such as biocompatibility, biodegradability and stimuli-responsiveness. These finding account for the large success of using hollow nanoparticles for drug delivery and biomedical imaging where their capacity for encapsulating sensitive materials (fluorescent markers, therapeutics, field-responsive agents…) has been widely exploited. Furthermore, multilayered capsules can be used to gain information on specific biological processes (nanobio interactions at different scale, Figure 9) depending on the particle characteristics, such as composition, shape, size, and surface functionality [89].

The above template may be assembled with several naturally derived polymers (e.g., polysaccharide) and be loaded with one or more hydrophobic drug/probe. Design of capsules containinga multi-compartment structure is a very powerful way to enrich their activity in bio-nanotechnology providing multiple functionality as for instance for therapy and diagnosis, i.e., theranostic.

Biocompatible and biodegradable nanocapsules could be useful for several reasons. (1) The hollow nanoparticle with a large inner space is able to encapsulate large quantity of therapeutic molecules, and may also protect them from degradation before delivered to target sites. (2) The small size of nanocapsules makes them compatible with several administration ways including intravenous injection and conferring a rapid absorption and release behavior provided by high abilities of their diffusion and volume change. (3) The biocompatibility and biodegradability from the biopolymers (i.e., building materials), make them less toxic to cells and organs.

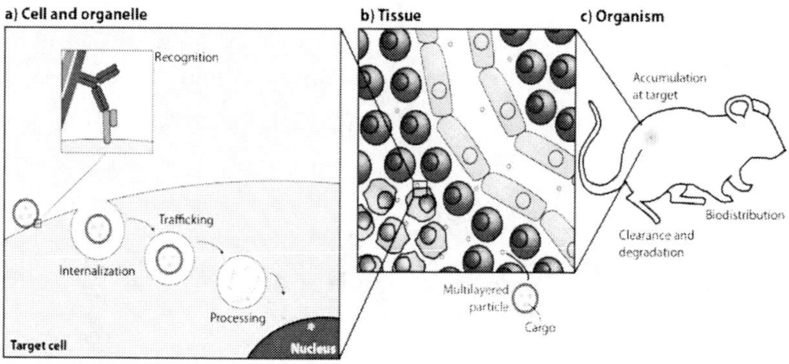

Figure 9. Interactions between multilayered particles with biological systems at various levels: (a) Interactions with cells and subcellular organelles involve the recognition of particles by target cells and subsequent intracellular processing for controlled release; (b) Interactions with tissue involve the accumulation and penetration of particles within a tissue to access the targeted cells; (c) Interactions at the organism level involve particle pharmacokinetics and biodistribution, including particle clearance and degradation [89].

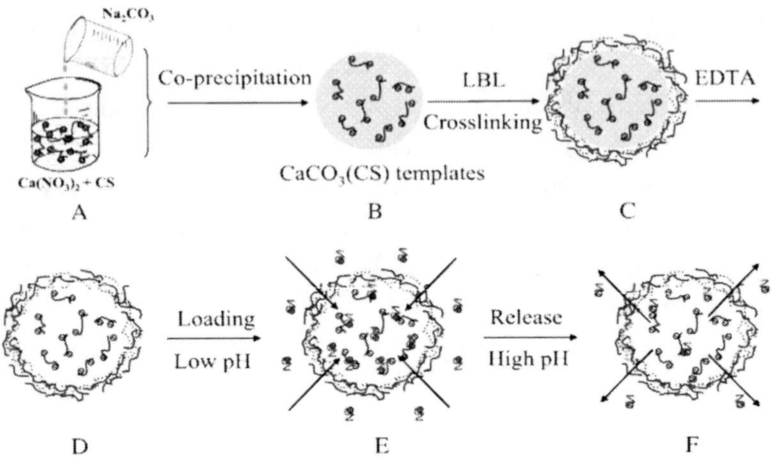

Figure 10. Schematic diagram for the preparation of CS-integrated multilayered microcapsules and related loading and release of FITC-BSA. A→B, Preparation of CS-integrated $CaCO_3$ templates. B→C, Self-assembly and cross-linking of multilayered shell on $CaCO_3$(CS) templates. C→D, Decomposition of the $CaCO_3$ templates and formation of CS-integrated multilayered microcapsules. D→E, Loading of BSA into CS-integrated microcapsules at a pH at which BSA has net positive charges. E→F, Release of BSA from CS-integrated microcapsules at neutral pH where BSA has net negative charges. CS, Cross-linked multilayered shell, BSA. The net charge of BSA is positive below its pI (pH 5.0) and negative above its pI [90].

In this section, judiciously selected examples from the literature will be presented, more precisely the major advantages and benefits of using nanocapsules as drug delivery systems, notably in terms of drug loading and targeting ability.

III.1. pH-Responsive Drug Delivery Nanocapsules

pH-responsive drug delivery systems are very attractive for intracellular drug delivery because of the pH fall between intracellular endosomes and lysosomes and physiological conditions. For instance, Zhao et al. reported biodegradable drug delivery multilayer polyelectrolyte capsules made of the chondroitin sulfate/poly(L-lysine) combination presenting pH-tunable loading and release properties [90]. Bovine serum albumin labeled with fluorescein isothiocyanate (FITC-BSA) was entrapped as a model drug. The authors evidenced that the loading is enhanced at a pH value near the isoelectric point of BSA (pH 5.0) and released faster at higher pH (Figure 10).

De Geest et al. discussed poly-L arginine/dextran sulfate multilayer capsules presenting pH responsive behavior [91]. Interestingly, the release is applicable to a specific environmental pH in the target tissues or cells (e.g., stomach (pH 1–2); intestine (pH 8.4); or endosome (pH 6.0–6.5).

III.2. Thermo-Responsive Drug Delivery Nanocapsules

Zhang et al. designed hollow core-shell structure nanocapsules composed of Pluronic®F127 and chitosan for the controlled intracellular release of low molecular weight compounds (ethidium bromide) [83]. The thermal responsiveness of the nanocapsules was exploited via the thermal from 37 nm at 37°C to 240 nm at 4°C. This phenomenon was accompanied by a change in permeability. In the swollen state, i.e., at low temperature, the wall capsules' permeability increased allowing free diffusion of the encapsulated molecules. Intracellular delivery was achieved in cytosol of both cancerous and non-cancerous mammalian cells. In another implementation the same group of authors reported the controlled intracellular release of trehalose, a highly hydrophilic molecule, i.e., a glucose disaccharide [92]. A major result of the study was that cytotoxicity of the nanocapsules was found negligible. Thermally sensitive cationic polymer nanocapsules were developed for the specific delivery of cytosolic in combination with the efficient gene silencing of siRNA [93]. Nanocapsules based on pluronic/poly(ethylenimine) exhibited a thermally reversible swelling/deswelling volume expansion behavior. In the collapsed state, nanocapsules had an average size of 118.9 +/- 15.3 nm at 37°C, while they assumed size of 412.3 +/- 83.2 nm at 15°C in the swollen state. siRNA-PEG conjugate was loaded through electrostatic interactions in the collapsed pluronic®/PEI2K nanocapsules taking advantage of the positive value of the surface zeta potential. The so-designed siRNA-PEG/nanocapsules were then transfected to the cells at 37°C for efficient cellular uptake. Abrupt volume expansion was induced in the transfected cells with a brief cold shock providing physical burst out of the endosomal membrane.

III.3. Redox-Responsive Drug Delivery Nanocapsules

It is well-known that a significant difference in redox potential exists between the extracellular and intracellular compartments. This change is correlated to the higher glutathione (GSH, reducing agent) concentrations in cytosol than in the extracellular fluids [94]. Hollow particles that can be subjected to reduction by GSH are then very promising and attractive for controlled intracellular drug delivery [95, 96]. Zheng et al. designed and fabricated novel redox-responsive and biodegradable submicron capsules through the layer-by-layer technique using chitosan and poly(L-aspartic acid) for transmucosal delivery of peptides and proteins (Figure 11) [97]. Insulin could be efficiently entrapped within the nanocapsules, and the amount of released insulin could be regulated by changing the GSH level. The cell viability test evidenced that all types of submicron capsules had no cytotoxicity and good cytocompatibility.

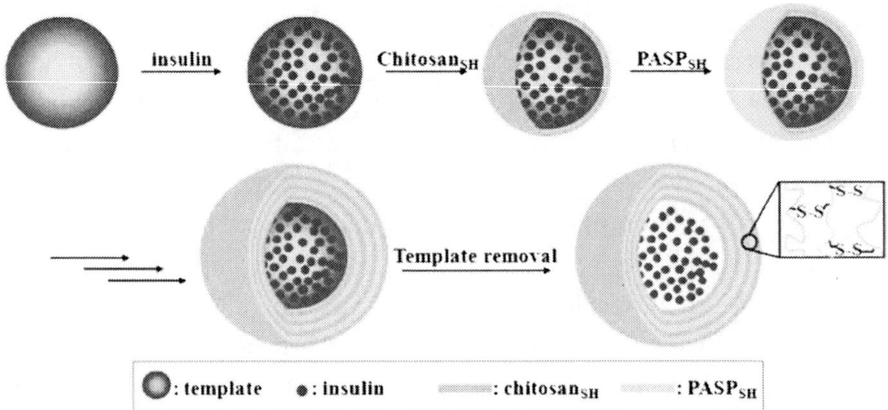

Figure 11. Formation of submicrocapsule *via* Layer-by-layer assembly of thiol-modified poly(L-aspartic acid) (PASPSH) and chitosan (CSSH) on silica template [97].

III.4. Light-Responsive Drug Delivery Nanocapsules

The photo-control of host-guest interaction between azobenzene and beta-cyclodextrin was presented as a new approach to prepare poly((itaconoyloxy)ethyl methacrylate)block-poly(N-isopropylacrylamide) -based hollow nanospheres [98]. The advantages of the method are that it is simple, environmentally friendly, and highly efficient. The so-obtained hollow nanospheres exhibited a typical "breathing" behavior, enabling the controlled release of doxorubicin hydrochloride. Cellular toxicity was evaluated and evidenced the biocompatibility feature of the hollow nanospheres making them appealing candidates as promising doxorubicin hydrochloride carriers.

Conclusion

This chapter highlights the diversity of methods that have been implemented to fabricate "smart" hollow nanoparticles (nanocapsules) from biopolymers, as well as some applications of these nanomaterials in the fields of biology and nanomedecine. As such, hollow nanoparticles with varied chemical compositions and exhibiting a large panel of physico-chemical properties can successfully be elaborated. However, the commercialization of this process is still very difficult due to the processing time, the complexity of fabrication process and operation cost. Future development of the fabrication methods is still essential to expand the application of the particles.

References

[1] Wagner, V., et al., The emerging nanomedicine landscape. *Nat Biotechnol,* 2006. 24(10): p. 1211-7.

[2] Bozzuto, G. and Molinari, A. Liposomes as nanomedical devices. *Int J Nanomedicine,* 2015. 10: p. 975-99.

[3] Singh, R. and Lillard, J.W., Nanoparticle-based targeted drug delivery. *Experimental and molecular pathology,* 2009. 86(3): p. 215-223.

[4] Caruso, F., Hollow capsule processing through colloidal templating and self-assembly. *Chemistry-A European Journal,* 2000. 6(3): p. 413-419.

[5] Peyratout, C.S. and Dahne, L., Tailor-made polyelectrolyte microcapsules: from multilayers to smart containers. *Angew Chem Int Ed Engl,* 2004. 43(29): p. 3762-83.

[6] Marinakos, S.M., et al., Gold particles as templates for the synthesis of hollow polymer capsules. Control of capsule dimensions and guest encapsulation. *Journal of the American Chemical Society,* 1999. 121(37): p. 8518-8522.

[7] Wu, M., et al., Synthesis of nanometer-sized hollow polymer capsules from alkanethiol-coated gold particles. *Chemical Communications,* 2000(9): p. 775-776.

[8] Caruso, F., R.A. Caruso, and Möhwald, H., Nanoengineering of inorganic and hybrid hollow spheres by colloidal templating. *Science,* 1998. 282(5391): p. 1111-1114.

[9] De Koker, S., Hoogenboom, R. and De Geest, B.G., Polymeric multilayer capsules for drug delivery. *Chem Soc Rev,* 2012. 41(7): p. 2867-84.

[10] Tong, W., Song, X. and Gao, C., Layer-by-layer assembly of microcapsules and their biomedical applications. *Chem Soc Rev,* 2012. 41(18): p. 6103-24.

[11] Johnston, A.P.R., et al., Layer-by-layer engineered capsules and their applications. *Current Opinion in Colloid & Interface Science,* 2006. 11(4): p. 203-209.

[12] Jones, O.G. and McClements, D.J., Recent progress in biopolymer nanoparticle and microparticle formation by heat-treating electrostatic protein-polysaccharide complexes. *Adv Colloid Interface Sci,* 2011. 167(1-2): p. 49-62.

[13] Nitta, S.K. and Numata, K., Biopolymer-Based Nanoparticles for Drug/Gene Delivery and Tissue Engineering. *International Journal of Molecular Sciences,* 2013. 14(1): p. 1629-1654.
[14] Chen, K.J., et al., A thermoresponsive bubble-generating liposomal system for triggering localized extracellular drug delivery. *ACS Nano,* 2013. 7(1): p. 438-46.
[15] You, J., et al., Effective photothermal chemotherapy using doxorubicin-loaded gold nanospheres that target EphB4 receptors in tumors. *Cancer Res,* 2012. 72(18): p. 4777-86.
[16] Zhang, L., et al., General route to multifunctional uniform yolk/mesoporous silica shell nanocapsules: a platform for simultaneous cancer-targeted imaging and magnetically guided drug delivery. *Chemistry,* 2012. 18(39): p. 12512-21.
[17] Schroeder, A., et al., Ultrasound triggered release of cisplatin from liposomes in murine tumors. *J Control Release,* 2009. 137(1): p. 63-8.
[18] Deng, Z., et al., Hollow chitosan-silica nanospheres as pH-sensitive targeted delivery carriers in breast cancer therapy. *Biomaterials,* 2011. 32(21): p. 4976-86.
[19] Koo, A.N., et al., Disulfide-cross-linked PEG-poly(amino acid)s copolymer micelles for glutathione-mediated intracellular drug delivery. *Chem Commun (Camb),* 2008(48): p. 6570-2.
[20] Gullotti, E., Park, J. and Yeo, Y., Polydopamine-based surface modification for the development of peritumorally activatable nanoparticles. *Pharm Res,* 2013. 30(8): p. 1956-67.
[21] Gero, D. and Jong-Dal, H., Buildup of ultrathin multilayer films by a self-assembly process, 1 consecutive adsorption of anionic and cationic bipolar amphiphiles on charged surfaces. *Makromolekulare Chemie. Macromolecular Symposia,* 1991. 46(1): p. 321-327.
[22] Decher, G., Fuzzy nanoassemblies: toward layered polymeric multicomposites. *Science,* 1997. 277(5330): p. 1232-1237.
[23] Lvov, Y., Decher, G. and Sukhorukov, G., Assembly of thin films by means of successive deposition of alternate layers of DNA and poly (allylamine). *Macromolecules,* 1993. 26(20): p. 5396-5399.

[24] Lvov, Y., et al., Alternate Assembly of Ordered Multilayers of SiO2 and Other Nanoparticles and Polyions. *Langmuir,* 1997. 13(23): p. 6195-6203.

[25] Wagberg, L., et al., The build-up of polyelectrolyte multilayers of microfibrillated cellulose and cationic polyelectrolytes. *Langmuir,* 2008. 24(3): p. 784-95.

[26] Zhao, Q., et al., Assembly of multilayer microcapsules on CacO3 particles from biocompatible polysaccharides. *J Biomater Sci Polym* Ed, 2006. 17(9): p. 997-1014.

[27] Borges, J.O. and Mano, J.F. Molecular interactions driving the layer-by-layer assembly of multilayers. *Chemical reviews,* 2014. 114(18): p. 8883-8942.

[28] Gil, P.R., et al., Nanoparticle-modified polyelectrolyte capsules. *Nano Today,* 2008. 3(3): p. 12-21.

[29] Yang, X., et al., Polyelectrolyte and molecular host ion self-assembly to multilayer thin films: An approach to thin film chemical sensors. *Sensors and Actuators B: Chemical,* 1997. 45(2): p. 87-92.

[30] Belbekhouche, S., Mansour, O. and Carbonnier, B., Promising sub-100 nm tailor made hollow chitosan/poly (acrylic acid) nanocapsules for antibiotic therapy. *Journal of colloid and interface science,* 2018. 522: p. 183-190.

[31] Decher, G. and Hong, J., *Thin Solid Films,* 210–211, 831;(c) Decher. G.(1997) *Science,* 1992. 277: p. 1232.

[32] Caruso, F., et al., 1. Ultrathin multilayer polyelectrolyte films on gold: construction and thickness determination. *Langmuir,* 1997. 13(13): p. 3422-3426.

[33] Ladam, G., et al., In situ determination of the structural properties of initially deposited polyelectrolyte multilayers. *Langmuir,* 2000. 16(3): p. 1249-1255.

[34] Elbert, D.L., Herbert, C.B. and Hubbell, J.A., Thin polymer layers formed by polyelectrolyte multilayer techniques on biological surfaces. *Langmuir,* 1999. 15(16): p. 5355-5362.

[35] Ruths, J., et al., Polyelectrolytes I: polyanion/polycation multilayers at the air/monolayer/water interface as elements for quantitative

polymer adsorption studies and preparation of hetero-superlattices on solid surfaces. *Langmuir,* 2000. 16(23): p. 8871-8878.

[36] Picart, C., et al., Buildup mechanism for poly (L-lysine)/hyaluronic acid films onto a solid surface. *Langmuir,* 2001. 17(23): p. 7414-7424.

[37] Pardo-Yissar, V., et al., Layered polyelectrolyte films on Au electrodes: characterization of electron-transfer features at the charged polymer interface and application for selective redox reactions. *Langmuir,* 2001. 17(4): p. 1110-1118.

[38] Lavalle, P., et al., Comparison of the structure of polyelectrolyte multilayer films exhibiting a linear and an exponential growth regime: an in situ atomic force microscopy study. *Macromolecules,* 2002. 35(11): p. 4458-4465.

[39] Boulmedais, F., et al., Buildup of exponentially growing multilayer polypeptide films with internal secondary structure. *Langmuir,* 2003. 19(2): p. 440-445.

[40] Halthur, T.J. and Elofsson, U.M., Multilayers of charged polypeptides as studied by in situ ellipsometry and quartz crystal microbalance with dissipation. *Langmuir,* 2004. 20(5): p. 1739-1745.

[41] Richert, L., et al., Layer by Layer Buildup of Polysaccharide Films: Physical Chemistry and Cellular Adhesion Aspects. *Langmuir,* 2004. 20(2): p. 448-458.

[42] Yuan, W. and Li, C.M., Exponentially growing layer-by-layer assembly to fabricate pH-responsive hierarchical nanoporous polymeric film and its superior controlled release performance. *Chemical Communications,* 2010. 46(48): p. 9161-9163.

[43] Picart, C., et al., Molecular basis for the explanation of the exponential growth of polyelectrolyte multilayers. *Proc Natl Acad Sci U S A,* 2002. 99(20): p. 12531-5.

[44] Salomaki, M. and Kankare, J., Modeling the growth processes of polyelectrolyte multilayers using a quartz crystal resonator. *J Phys Chem* B, 2007. 111(29): p. 8509-19.

[45] Hoda, N. and Larson, R.G., Modeling the buildup of exponentially growing polyelectrolyte multilayer films. *J Phys Chem* B, 2009. 113(13): p. 4232-41.

[46] Dubas, S.T. and Schlenoff, J.B., Factors controlling the growth of polyelectrolyte multilayers. *Macromolecules,* 1999. 32(24): p. 8153-8160.

[47] Büscher, K., et al., Influence of adsorption conditions on the structure of polyelectrolyte multilayers. *Langmuir,* 2002. 18(9): p. 3585-3591.

[48] Tan, H.L., et al., Temperature dependence of polyelectrolyte multilayer assembly. *Langmuir,* 2003. 19(22): p. 9311-9314.

[49] Klitzing, R.v., et al., Short range interactions in polyelectrolyte multilayers. *Current opinion in colloid & interface science,* 2004. 9(1-2): p. 158-162.

[50] Sukhorukov, G.B., et al., Layer-by-layer self assembly of polyelectrolytes on colloidal particles. *Colloids and Surfaces A: physicochemical and engineering aspects,* 1998. 137(1-3): p. 253-266.

[51] Mulvaney, P., Surface plasmon spectroscopy of nanosized metal particles. *Langmuir,* 1996. 12(3): p. 788-800.

[52] Schneider, G. and Decher, G., Functional core/shell nanoparticles via layer-by-layer assembly. Investigation of the experimental parameters for controlling particle aggregation and for enhancing dispersion stability. *Langmuir,* 2008. 24(5): p. 1778-1789.

[53] Gittins, D.I. and Caruso, F., Tailoring the Polyelectrolyte Coating of Metal Nanoparticles. *The Journal of Physical Chemistry B,* 2001. 105(29): p. 6846-6852.

[54] Bhattacharjee, R.R., Chakraborty, M. and Mandal, T.K., Synthesis of dendrimer-stabilized gold-polypyrrole core-shell nanoparticles. *Journal of nanoscience and nanotechnology,* 2003. 3(6): p. 487-491.

[55] Gittins, D.I. and Caruso, F., Multilayered polymer nanocapsules derived from gold nanoparticle templates. *Advanced Materials,* 2000. 12(24): p. 1947-1949.

[56] Schneider, G. and Decher, G., From Functional Core/Shell Nanoparticles Prepared via Layer-by-Layer Deposition to Empty Nanospheres. *Nano Letters,* 2004. 4(10): p. 1833-1839.

[57] Becker, A.L., Johnston, A.P. and Caruso, F., Layer-by-layer-assembled capsules and films for therapeutic delivery. *Small,* 2010. 6(17): p. 1836-52.

[58] Yu, L.W. and Hartwick, R.A., *J. Chromatogr. Sci.,* 1989. 27: p. 176.

[59] Belbekhouche, S., et al., Glucose-sensitive polyelectrolyte microcapsules based on (alginate/chitosan) pair. *Carbohydrate Polymers,* 2018. 184: p. 144-153.

[60] Nagaraja, A.T., Pradhan, S. and McShane, M.J., Poly (vinylsulfonic acid) assisted synthesis of aqueous solution stable vaterite calcium carbonate nanoparticles. *J Colloid Interface Sci,* 2014. 418: p. 366-72.

[61] Biswas, A., Nagaraja, A.T. and McShane, M.J., Fabrication of Nanocapsule Carriers from Multilayer-Coated Vaterite Calcium Carbonate Nanoparticles. *ACS Applied Materials & Interfaces,* 2014. 6(23): p. 21193-21201.

[62] Bunton, C.A., et al., Ion binding and reactivity at charged aqueous interfaces. *Accounts of Chemical Research,* 1991. 24(12): p. 357-364.

[63] Cuomo, F., et al., Vesicle-Templated Layer-by-Layer Assembly for the Production of Nanocapsules. *Langmuir,* 2010. 26(13): p. 10555-10560.

[64] Cuomo, F., et al., pH-responsive liposome-templated polyelectrolyte nanocapsules. *Soft Matter,* 2012. 8(16): p. 4415-4420.

[65] Rahn-Chique, K., et al., Nanoemulsion stability: Experimental evaluation of the flocculation rate from turbidity measurements. *Advances in Colloid and Interface Science,* 2012. 178: p. 1-20.

[66] Ni, Y., et al., Synthesis, characterization and properties of hollow nickel phosphide nanospheres. *Nanotechnology,* 2006. 17(19): p. 5013.

[67] Zimmermann, C., et al., Nanoscale gold hollow spheres through a microemulsion approach. *Small,* 2007. 3(8): p. 1347-1349.

[68] Buchold, D.H. and Feldmann, C. Nanoscale γ-AlO (OH) hollow spheres: synthesis and container-type functionality. *Nano letters,* 2007. 7(11): p. 3489-3492.

[69] Li, Y., et al., Hollow spheres of mesoporous aluminosilicate with a three-dimensional pore network and extraordinarily high hydrothermal stability. *Nano Letters,* 2003. 3(5): p. 609-612.

[70] Abbas, S., et al., Fabrication of polymeric nanocapsules from curcumin-loaded nanoemulsion templates by self-assembly. *Ultrasonics sonochemistry,* 2015. 23: p. 81-92.

[71] Beaman, D.K., Robertson, E.J. and Richmond, G.L. Unique Assembly of Charged Polymers at the Oil− Water Interface. *Langmuir,* 2011. 27(6): p. 2104-2106.

[72] Li, Y., et al., Controlling lipid nanoemulsion digestion using nanolaminated biopolymer coatings. *Journal of microencapsulation,* 2011. 28(3): p. 166-175.

[73] Preetz, C., et al., Preparation and characterization of biocompatible oil-loaded polyelectrolyte nanocapsules. *Nanomedicine: Nanotechnology, Biology and Medicine, 2008*. 4(2): p. 106-114.

[74] Szczepanowicz, K., et al., Encapsulation of liquid cores by layer-by-layer adsorption of polyelectrolytes. *Journal of microencapsulation,* 2010. 27(3): p. 198-204.

[75] Holme, M.N., et al., Shear-stress sensitive lenticular vesicles for targeted drug delivery. *Nature nanotechnology,* 2012. 7(8): p. 536-543.

[76] Zhang, J., et al., Ultrasound-Triggered Drug Delivery for Breast Tumor Therapy Through iRGD-Targeted Paclitaxel-Loaded Liposome-Microbubble Complexes. *Journal of biomedical nanotechnology,* 2018. 14(8): p. 1384-1395.

[77] Zhao, Q., et al., Hyaluronic acid and carbon dots-gated hollow mesoporous silica for redox and enzyme-triggered targeted drug delivery and bioimaging. *Materials Science and Engineering: C,* 2017. 78: p. 475-484.

[78] Jayasundar, R. and Singh, V., In vivo temperature measurements in brain tumors using proton MR spectroscopy. *Neurol India,* 2002. 50(4): p. 436-439.

[79] Issels, R.D., Hyperthermia adds to chemotherapy. *European journal of cancer,* 2008. 44(17): p. 2546-2554.

[80] Fu, Q., et al., Thermoresponsive transport through ordered mesoporous silica/PNIPAAm copolymer membranes and microspheres. *Langmuir,* 2007. 23(1): p. 170-174.

[81] Schild, H.G., Poly (N-isopropylacrylamide): experiment, theory and application. *Progress in polymer science,* 1992. 17(2): p. 163-249.

[82] Chen, S., et al., Temperature-responsive magnetite/PEO– PPO–PEO block copolymer nanoparticles for controlled drug targeting delivery. *Langmuir,* 2007. 23(25): p. 12669-12676.

[83] Zhang, W., et al., Synthesis and characterization of thermally responsive pluronic F127– chitosan nanocapsules for controlled release and intracellular delivery of small molecules. *ACS nano,* 2010. 4(11): p. 6747-6759.

[84] Fomina, N., Sankaranarayanan, J. and Almutairi, A. Photochemical mechanisms of light-triggered release from nanocarriers. *Advanced drug delivery reviews,* 2012. 64(11): p. 1005-1020.

[85] Sankaranarayanan, J., et al., Multiresponse strategies to modulate burst degradation and release from nanoparticles. *Acs Nano,* 2010. 4(10): p. 5930-5936.

[86] Singh, R. and Whitesides, G.M., *Thiol-disulfide interchange.* 1993, John Wiley & Sons, Inc.: Chichester, UK. p. 633-658.

[87] Markovic, J., et al., Role of glutathione in cell nucleus. *Free radical research,* 2010. 44(7): p. 721-733.

[88] Kim, H.-C., et al., Thiol-responsive gemini poly (ethylene glycol)-poly (lactide) with a cystine disulfide spacer as an intracellular drug delivery nanocarrier. *Colloids and Surfaces B: Biointerfaces,* 2015. 127: p. 206-212.

[89] Yan, Y., Björnmalm, M. and Caruso, F., Assembly of layer-by-layer particles and their interactions with biological systems. *Chemistry of Materials,* 2013. 26(1): p. 452-460.

[90] Zhao, Q. and Li, B., pH-controlled drug loading and release from biodegradable microcapsules. *Nanomedicine: Nanotechnology, Biology and Medicine,* 2008. 4(4): p. 302-310.

[91] De Geest, B.G., et al., Intracellularly degradable polyelectrolyte microcapsules. *Advanced materials,* 2006. 18(8): p. 1005-1009.

[92] Zhang, W., et al., The encapsulation and intracellular delivery of trehalose using a thermally responsive nanocapsule. *Nanotechnology,* 2009. 20(27): p. 275101.

[93] Lee, S.H., et al., Thermally sensitive cationic polymer nanocapsules for specific cytosolic delivery and efficient gene silencing of siRNA: swelling induced physical disruption of endosome by cold shock. *Journal of Controlled Release,* 2008. 125(1): p. 25-32.

[94] Schafer, F.Q. and Buettner, G.R., Redox environment of the cell as viewed through the redox state of the glutathione disulfide/ glutathione couple. *Free Radical Biology and Medicine,* 2001. 30(11): p. 1191-1212.

[95] Saito, G., Swanson, J.A. and Lee, K.-D., Drug delivery strategy utilizing conjugation via reversible disulfide linkages: role and site of cellular reducing activities. *Advanced drug delivery reviews,* 2003. 55(2): p. 199-215.

[96] Qian, J. and Wu, F., Thermosensitive PNIPAM semi-hollow spheres for controlled drug release. *Journal of Materials Chemistry* B, 2013. 1(28): p. 3464-3469.

[97] Zheng, C., et al., Biodegradable and redox-responsive chitosan/poly (L-aspartic acid) submicron capsules for transmucosal delivery of proteins and peptides. *Journal of Materials Science: Materials in Medicine,* 2013. 24(4): p. 931-939.

[98] Zhang, H., et al., Photo-controlled host–guest interaction as a new strategy to improve the preparation of "breathing" hollow polymer nanospheres for controlled drug delivery. *Journal of Materials Chemistry* B, 2015. 3(43): p. 8528-8536.

In: Polymer Surfaces
Editor: Valente Erasmo Silva
ISBN: 978-1-53614-309-6

Chapter 3

RAPID FABRICATION OF AN ANTIREFLECTIVE PYRAMID STRUCTURE ON POLYSTYRENE SURFACES USED AS THE PROTECTIVE LAYER OF A SOLAR CELL

Han-Xiong Huang*[*] *and Yu-Jiang Peng
Lab for Micro/Nano Molding and Polymer Rheology,
The Key Laboratory of Polymer Processing Engineering
of the Ministryof Education, South China University of Technology,
Guangzhou, People's Republic of China

ABSTRACT

A promising strategy is proposed for facilely and successively replicating randomly arranged pyramids on an etched silicon wafer to polystyrene (PS) surface. First, a nickel mold insert with negative pyramids is fabricated via combining electroless plating and subsequent electroplating with an etched silicon wafer as a template. Then, by using microinjection compression molding with the nickel mold insert as a

[*] Corresponding Author Email: mmhuang@scut.edu.cn.

template, the pyramids on the insert are accurately transcribed to the PS surface. The pyramids on the PS replica surface have the hemlines of about 1–7 μm and heights of about 1–4 μm. The PS replica surface exhibits a reflectance of about 5% in a wavelength range of 400–900 nm and an enhanced transmittance. That is, the surface exhibits an enhanced light trapping effect, which originates from a combination of staggered arrangement and dense distribution of the pyramids with arc-shaped top and different sizes. Especially, a thin film solar cell covered with the PS replica exhibits increased power conversion efficiency enhancement of up to 7.9% under normal incidence comparing to that covered with the PS counterpart. The solar cell covered with the PS replica also shows an excellent photovoltaic performance over a wide range of incident angles (0–70°). The proposed fast and efficient replication strategy can be an excellent candidate for developing antireflective protective layers without complicated procedures and expensive materials.

Keywords: light trapping effect, antireflection, microstructured surface, pyramid, solar cell, microinjection compression molding

INTRODUCTION

It is crucial to suppress reflection of light and improve its transmission in developing high-performance photovoltaic devices, especially solar cells [1]. Antireflection coating and surface texturing are two main methods for reducing reflection losses at air-substrate interface and imparting a surface with antireflective property [2]. The two methods can increase the short circuit current and boost the conversion efficiency of solar cells [3]. Compared to antireflection coatings, texturized surface exhibits better durability and mechanical stability due to its monolithic structure and homogeneous material. Many efforts have been devoted to fabricate the structured antireflective surfaces possessing quasi-omnidirectional, broadband antireflective properties. Different micro/nanostructured arrays, such as nanopillar [4, 5], nanocone [6], micropyramid structure [7–9], V-groove [10], nanowire [11, 12], and micro-lenses [13, 14], have been artificially created. Among the antireflective structures developed, pyramid structure has been intensively applied in light trapping to increase the conversion efficiency of solar cell through its light trapping effect [15–17]. Texturing pyramids on silicon wafers by alkaline etching is an established technique. However, this approach is not suitable for thin film solar cells where the thickness of active region is only several microns or even hundreds of nanometers. Moreover, as an essential element of solar cell for protecting it from external shocks and corrosion [18], the protective layer also reflects a fraction of incident light. In practical applications, glass or polymer based protective layer reflects normal incident light from 4% to more than 6.5% from air-substrate interface [19]. Reflection losses from protective layer are notably disadvantageous because the conversion efficiency of solar cells relies on efficiently transmitted energy.

Although glass and metals are still the most commonly used as top- and bottom-sealing substrates for solar cells, great efforts have been expended to develop polymeric and biopolymeric supporting substrates [20–26] for solar cells. Griffini et al. [20] developed an organic-inorganic hybrid coating material as luminescent downshifting host matrix, simultaneously enhancing the efficiency and durability of flexible organic photovoltaic devices. Especially, polymeric thin layers with patterned surfaces can impart solar cells with multifunctional properties when acting as their supporting substrates. Patterning micro/nanostructures on the external side of polymer-based solar cells results in their hydrophobic/ superhydrophobic properties and so self-cleaning effect [21, 23–25]. More interestingly, using structured surfaces of polymeric protective layer to enhance light absorption [22–26] is a vogue topic nowadays. Polymeric and biopolymeric materials can be used in electrolytes and photoanodes other than supporting substrates for solar cells [27, 28].

Polymers with high transmittance are recently emerging as a favorable alternative especially for flexible solar cells, due to their flexibility and affordable cost. Replacing glass with polymeric layer can provide a more durable mechanical protective layer and eliminate glass breakage. There are a few techniques reported for fabricating polymer-based protective layers that can suppress surface reflection, such as hot embossing [24], UV imprint [23], and sol-gel process [22, 25]. Lee et al. [24] fabricated a solar cell protective film with a moth-eye anti-reflection structure by hot-embossing to suppress the reflection of incident light, resulting in a total increase of 3.9% in accumulated electric energy of the solar cell. In hot-embossing process, polymer is heated to above its glass transition temperature yet below its melting or flow temperature, and the replication is achieved using a combination of heat and pressure, spending tens of minutes. The UV imprint process comprises spin-coating and UV exposure steps and so is time-consuming and difficult to separate the film from the mold. Kang et al. [25] fabricated a polydimethylsiloxane (PDMS) film with micropyramids on its surface by sol-gel process. The result demonstrated that power conversion efficiency is increased by 6.8% when using the PDMS film as a protective layer of solar cell, indicating that the pyramid structure exhibits an anti-reflection property. However, it takes even hours before the PDMS film is completely cured in sol-gel process and a cautious separation for the PDMS film and mold is needed. UV imprint and sol-gel processes are capable of fabricating micro/nanostructures on the polymer substrate surface. However, when vacuum is not used during both processes, air may be trapped in the cavity, resulting in bubble defects [29, 30]. Therefore, sufficient time must be provided for the air to escape. Injection molding can provide a faster and cheaper alternative. Injection molding is an industrially established technique and capable of replicating micro/nanostructures on polymer surfaces [31–33]. Nevertheless, to the best knowledge of authors, only few researches were reported on using injection molding to prepare antireflective micro/nanostructured surfaces [34, 35], especially micropyramid structured surfaces.

Based on the aforementioned background, a strategy that combines electroless plating, electroplating, and microinjection compression molding (μ-ICM) is proposed to fabricate antireflective microstructured surface in this work. Using the proposed strategy, PS replicas with randomly arranged micropyramids on their surfaces are fabricated, which exhibit an antireflective property and enhanced light transmittance.

EXPERIMENTAL

The overall process for fabricating the PS replicas with pyramids from etched silicon wafer consisted of three steps, that is, silicon wafer etching (Figure 1(1)), nickel mold insert fabrication (Figure 1(2)), and PS replica molding (Figure 1(3)).

Silicon Wafer Etching

Boron-doped (1–3 Ω cm) n-type, (100)-oriented silicon (Si) wafers with a thickness of 200 ± 20 μm were used. The wafers were cut into square plates with a dimension of 30 × 30 mm^2 (Figure 1(a)).

The plates were ultrasonically cleaned in acetone and then in deionized water for 15 min for each step, to entirely remove organics and other impurities on the wafers. Then the plates were anisotropically etched in aqueous alkaline solution (3 wt% NaOH, 10 wt% isopropyl alcohol) [10] at 80°C for 20 min to form the pyramids on their surfaces (Figure 1(b)). Finally, the etched silicon wafers were thoroughly rinsed with ethanol and deionized water, and dried in nitrogen gas stream.

Nickel Mold Insert Fabrication

The steps for fabricating the nickel mold insert with negative pyramid structures comprised electroless nickel plating, electroplating, and etching (Figure 1(2)) in this work. Electroless nickel plating can be used to deposit highly electrical conductivity nickel layer on microscale structures. Electroplating is an inexpensive way to replicate micro/nanostructures, and replicated nickel insert can be used as a template in injection molding.

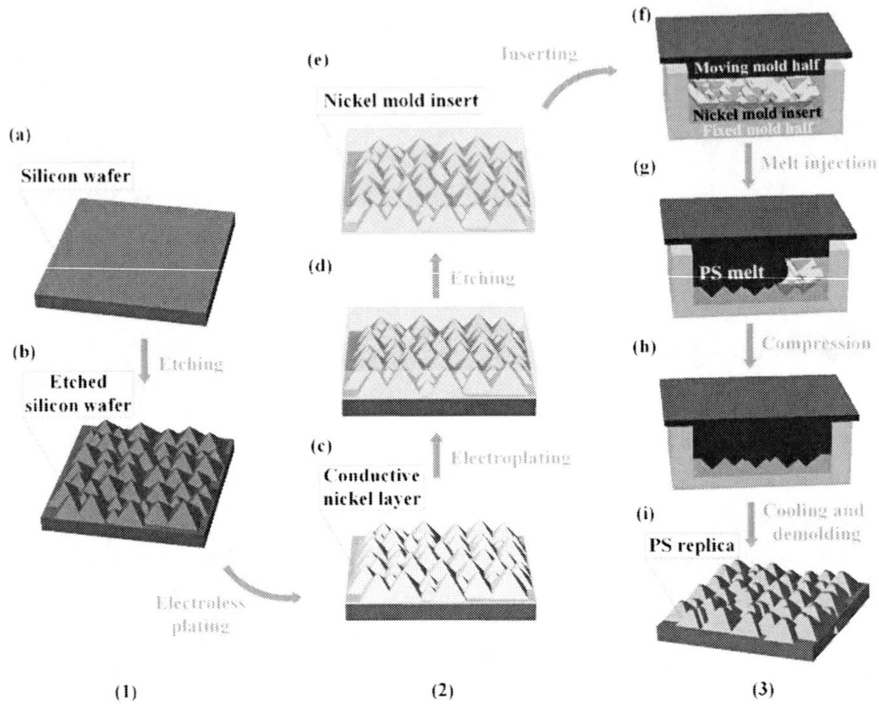

Figure 1. Schematics of overall fabrication process: (1) texturization of pyramids on silicon wafer by alkaline etching, (2) transfer of pyramids to nickel mold insert by electroless nickel plating and electroplating on surface of etched silicon, and (3) molding of PS replica with pyramid structure on surface by μ-ICM.

Table 1. Compositions and concentrations in solutions and process parameters used for electroless nickel plating

Step	Composition	Concentration	Parameter
Sensitization	$SnCl_2 \cdot H_2O$	10 g/L	25°C, 5 min
	HCl	40 mL/L	
Activation	$PdCl_2 \cdot H_2O$	1 g/L	25°C, 5 min
	HCl	10 mL/L	
Nickel deposition	$NiSO_4 \cdot 6H_2O$	30 g/L	50°C, 5 min, pH = 9.1
	$NaH_2PO_2 \cdot H_2O$	30 g/L	
	$Na_3C_6H_5O_7 \cdot 2H_2O$	20 g/L	

Table 2. Compositions and concentrations in solution and process parameters used for electroplating

Composition	Concentration
$Ni(NH_2SO_3)_2 \cdot 4H_2O$	300 g/L
$NiCl_2$	15 g/L
$C_{12}H_{25}SO_4Na$	0.5 g/L
H_3BO_3	35 g/L
50°C, pH = 3.9~4.0	

The electroless nickel plating on the aforementioned etched silicon wafer included pretreatment and nickel deposition, and the compositions and concentrations in the corresponding solutions and the process parameters are listed in Table 1. The etched silicon wafer was pretreated via sensitization and activation to provide catalytic nuclei for release of nickel ions in the reduction reaction of following electroless plating. Specifically, the silicon wafer was first immersed in the sensitization solution for 10 min and then in the activation solution for 5 min at room temperature. Subsequently, the wafer was dried in ambient atmosphere, ensuring that palladium catalytic nuclei were adhered to the surface of the pyramid structure. After that, the pretreated silicon wafer was immersed into the electroless nickel plating solution at 50°C with magnetic stirring. In the plating solution, $NiSO_4 \cdot 6H_2O$ and $NaH_2PO_2 \cdot H_2O$ acted as main salt and reductant, respectively. The solution was adjusted to pH = 9.1 using ammonia water. After 5 min of the electroless plating, the pyramid structured silicon wafer covered with conductive nickel layer (Figure 1(c)) was taken out from the solution and rinsed thoroughly with deionized water.

The electroplating was performed with the aforementioned electroless nickel plated silicon wafer acting as a cathode and a Ni-S plate as an anode in the electroplating solution. The compositions and concentrations in the solution and the process parameters are listed in Table 2. A power supply

(IT7000 DC Power Supply, ITECH Electronic Co. Ltd., USA) was used to provide a constant current density of 4 A/dm^2. To guarantee the uniformity of the electroplating solution, a magnetic stirring at 150 rpm was applied. After electroplating for 2 h, an approximately 400 μm thick nickel layer was deposited on the conductive nickel layer on the silicon wafer (Figure 1(d)). Finally, the silicon wafer was corroded away by wet etching with NaOH at 80°C, yielding the nickel mold insert after cleaning and drying (Figure 1(e)).

PS Replica Molding

The PS replicas were molded via the μ-ICM technology, as shown in Figure 1(3). PS (grade N1841H, Hong Kong Petrochemical Co. Ltd) with a refractive index of 1.6 was used as the material for molding the replica. The equipment used for molding the replicas comprised an 80 ton injection-molding machine (KM80SP180CX, Krauss-Maffei, Germany) and a microinjection compression mold equipped with a temperature control apparatus.

As a template, the previously fabricated nickel mold insert was mounted on the mold cavity surface, as shown in Figure 1(f). In the μ-ICM process, the mold cavity was first partially filled with molten PS when the mold was partially closed (Figure 1(g)). Then the mold was compressed in the thickness direction to finish melt filling (Figure 1(h)). Finally, the melt was packed under the compression force and simultaneously cooled down [38]. The molding parameters, including injection rate (154 cm^3/s), melt temperature (235°C), mold temperature (100°C), compression force (280 kN), and cooling time (14 s), were kept constant. It should be noted that the cycle time for molding the replicas was about 35 s. For comparison, PS counterparts with a smooth surface were molded per the same protocol without the nickel mold insert as the template.

Characterization

The surface topographies of the etched silicon wafers, prepared nickel mold inserts, and molded PS replicas were examined using a scanning electron microscopy (SEM; Merlin, Zeiss, Germany) at an accelerating voltage of 10 kV. For the PS replicas, the surfaces were coated with gold by ion sputtering prior to the SEM observation. The element compositions of the nickel mold inserts were analyzed using an energy dispersive spectroscopy (EDS; Inca Penta FET x3, Oxford Instruments, UK).

Optical reflectance and transmission spectra measurements on the etched silicon wafers and molded PS replicas were performed using an UV/VIS/NIR spectrophotometer (UV-3150, Shimadzu Corporation, Japan) in a wavelength range of 300–900 nm at an incidence angle (defined with respect to the surface normal) of 0°.

The molded PS replica with pyramid structures and PS counterpart were served as protective layer of the thin film solar cell. The photocurrent density–voltage (J-V) and external quantum efficiency (EQE) were characterized to compare the photovoltaic performance of the solar cell covered with the PS replica and PS counterpart.

An identical solar cell was used for all measurements to rule out the differences between different solar cells. The J-V measurements were performed using a solar simulator (SS-F5-3A, Enlitech Co. Ltd., Taiwan) under AM 1.5G illuminationsolar cell. A Keithley 2400 source meter was used to ramp the voltage from –0.1 to 2.7 V in a step of 0.02 V with a dwell time of 60 s to avoid heating of the solar cell. For angle-dependent photocurrents measurements, the distance between the solar cell and light source was kept constant. The EQE was measured by using a spectral response measurement system (QE-R, Enlitech Co. Ltd., Taiwan) in a wavelength range of 200–800 nm.

RESULTS AND DISCUSSION

Pyramids Formation

Figure 2(a, b) shows the SEM images of the etched silicon wafer surface. As can be seen, upright tetrahedral pyramids are randomly distributed and closely aligned throughout the surface, and almost each pyramid is partially overlapped with several adjacent pyramids of different sizes. The triangular pyramid surfaces are smooth. The hemlines and heights of the pyramids are roughly estimated to be 1–7 μm and 1–5 μm, respectively; their apex angles are about 75°.

The nickel mold insert was fabricated by means of atom by atom depositing the nickel layer on the surface of the etched silicon wafer via combining the electroless plating with electroplating in this work. It can be seen from Figure 2(c) that the pyramid surfaces are covered by a compact nickel layer with a thickness of about 1 μm (shown in the inset) after the electroless nickel plating. Figure 2(d) shows the SEM micrographs of the nickel mold insert surface after the nickel electroplating and removal of the etched silicon wafer. As can be seen, staggered negative pyramids of different sizes are randomly distributed on the surface of the mold insert. The hemlines of the pyramids are within the range of 1–7 μm. Hence, the pyramids on the etched silicon wafer are accurately replicated to the nickel mold insert with an inverse pattern.

The EDS test was performed to analyze the main chemical elements of the fabricated mold insert. It can be seen from the spectrum (Figure 3) of the nickel mold insert that the dominant peaks are nickel (Ni) and phosphorus (P), which demonstrates that the silicon substrate was thoroughly removed after the wet etching of the silicon with NaOH.

The PS replica with pyramids was obtained after cooling the PS melt in the μ-ICM process. The SEM micrographs of the PS replica surface are shown in Figure 2(e–f). As can be seen, the pyramids are randomly distributed throughout the PS replica surface and almost each pyramid is partially overlapped with several adjacent pyramids of different sizes. Comparing Figure 2(e, f) and Figure 2(a, b), the replicated pyramids on the

PS replica surface exhibit similar structure with the pyramids on the etched silicon wafer surface, and the main difference is arc-shaped top of the pyramids on the PS replica surface. As can be seen in Figure 2(e, f), the hemlines and heights of the pyramids on the PS replica surface are roughly estimated to be 1–7 μm and 1–4 μm, respectively; their trapezoidal surfaces and arc-shaped tops are much smooth. This indicates that the PS replicas are demolded without defect, which is ascribed to apex angle and low aspect-ratio of the pyramids and the smooth pyramid surfaces on the nickel mold insert (Figure 2(d)).

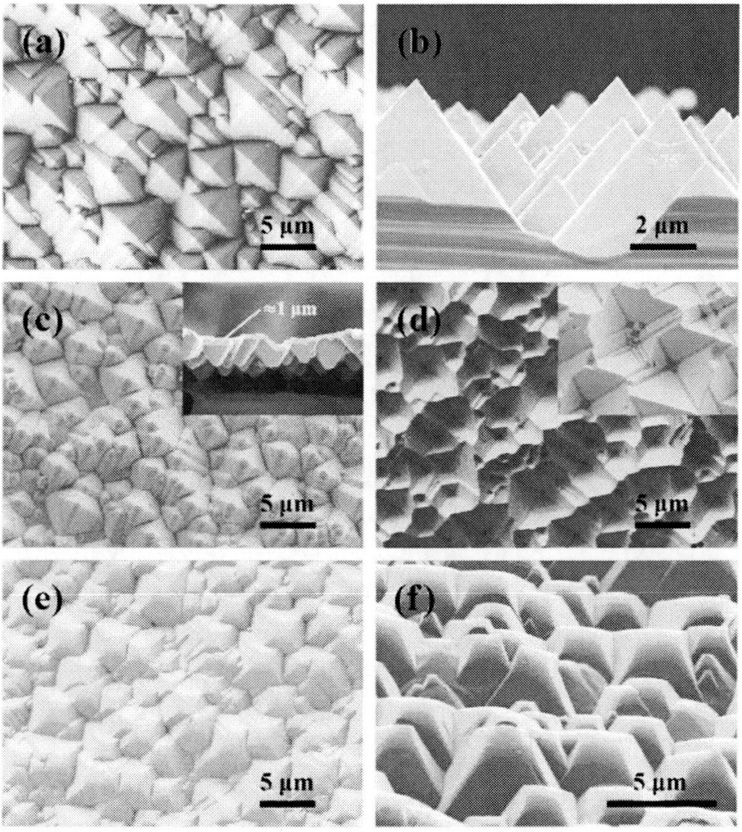

Figure 2. SEM images: (a) top and (b) lateral views of etched silicon wafer, (c) nickel layer on silicon wafer surface, inset is lateral view, (d) negative pyramids of nickel mold insert, inset is high magnification of negative pyramids, (e) top and (f) tilted views of PS replica molded by μ-ICM.

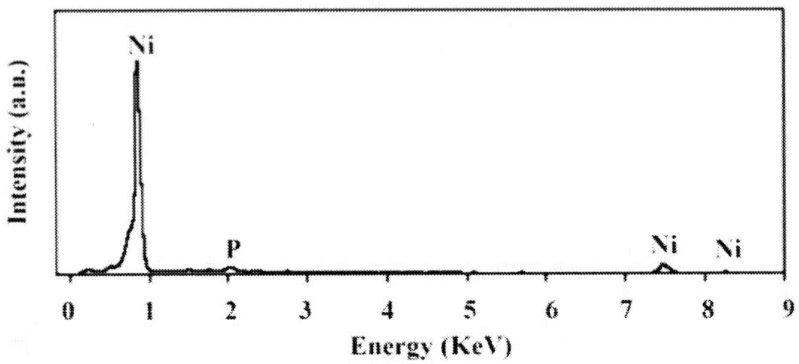

Figure 3. Energy dispersive spectroscopy spectrum of fabricated nickel mold insert.

Pyramids Effect on Light Trapping

The reflectances and transmittances of the fabricated PS replicas were investigated to characterize the optical properties of the pyramid structures. Figure 4(a) depicts the reflectance spectra of the as-received silicon wafer, etched silicon wafer, PS counterpart, and PS replica. With the pyramid structure, the reflectances of the PS replica and etched silicon wafer are considerably lower than those of the PS counterpart and silicon wafer, respectively, over the measured wavelength range. This light trapping effect can be solely ascribed to the influence of the pyramid structure and exclude any material-related effects.

Ideally, light trapping structures can not only reduce surface reflection losses, but also boost light transmittance. As shown in Figure 4(b), at the wavelength range of 400–900 nm, the PS counterpart exhibits an average transmittance of 89%, and the front side of the PS replica shows a substantially increased transmittance of 93%. For the PS replica, the back side exhibits merely one half of the transmittance for the front side, which can be ascribed to the fact that much of the light entering from its back side is totally internally reflected back to the interface between its back side and air. This means that when employed as a protective layer of solar cells, the PS replica can trap incident light that may normally be reflected and scattered back to air, thus further enhancing the power conversion efficiency of the solar cell.

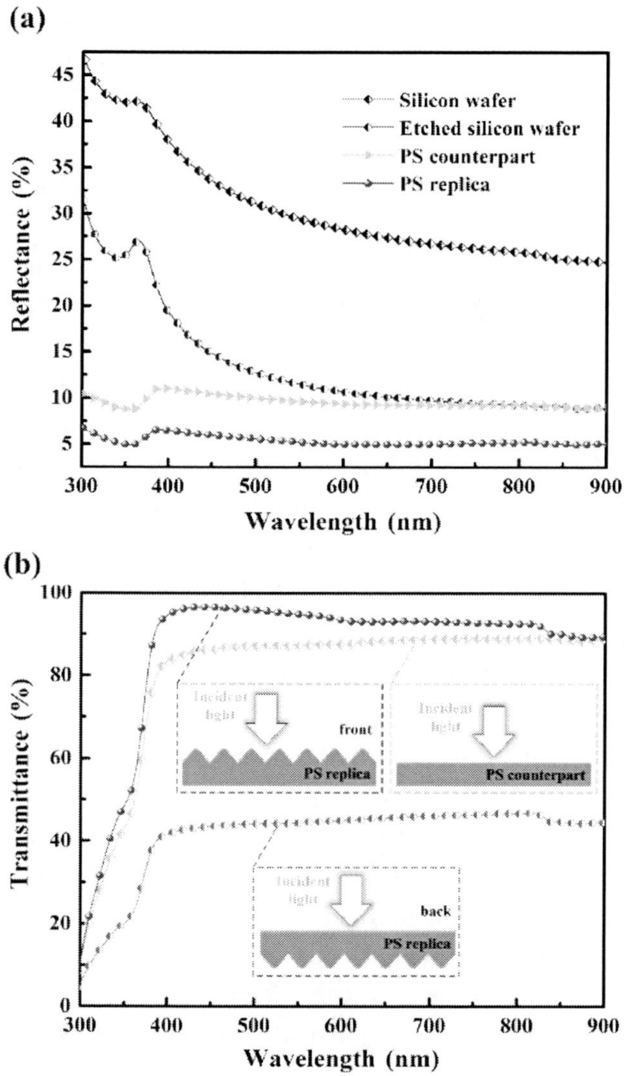

Figure 4. (a) Reflectance spectra of silicon wafer, etched silicon wafer, PS counterpart, and PS replica. (b) Transmittance spectra of PS counterpart and PS replica.

To visually display the antireflective property of the PS replica, Figure 5 (a) shows the photograph of the PS replica and PS counterpart exposed to a white fluorescent light. As compared to the PS counterpart, it is evident that the reflection from the PS replica is significantly suppressed and a light scattering phenomenon (i.e., multiple images) on the surface of the PS replica can be observed with the naked eyes. Figure 5 (b) shows a diffraction phenomenon of the PS replica using a laser with λ = 650 nm. It can be seen from Figure 5(b) that a nearly quarate spot is obtained from the light transmitted through the PS replica. The diffracted light pattern indicates high diffraction orders in the transmitted light, which result from the pyramids on the PS replica.

The aforementioned reduced reflectance and enhanced transmittance of the PS replica are analyzed from the following two aspects. On the one hand, pyramids can reflect incident light multiple times, resulting in an increased optical path length and thus a boosted absorption probability of light [37]. Figure 6 schematically illustrates the light trapping effect of pyramids. For incident light at different angles, a large fraction of the light transmits from air into the PS replica after its first pass, only small fraction of the light is reflected onto a neighboring pyramid. Then, most of the reflected light enter and is internally reflected within the replica after the second pass. Therefore, quasi-omnidirectionality (i.e., less angle-dependent of reflectance and transmittance characteristic) can be achieved in micropyramid structured surfaces.

On the other hand, the arrangement and dimension of pyramids play a crucial role in their light trapping effects. Comparing to the surface with uniform pyramids, the surface with staggered and differently-sized pyramids prepared in this work (Figure 2(e, f)) shows a much lower reflectance and enhanced transmittance [15, 16].

In addition, the arc-shaped top of the pyramid contributes to the suppression of reflection losses. Thus, the light trapping effect of the molded PS replica originates from the combination of the staggered arrangement and dense distribution of the pyramids with their different size and arc-shaped top (shown in Figure 2(e, f)). Consequently, transparent polymer layer prepared in this work can be used as a universal and efficient light collector, especially suitable to tackle the substantial reflection losses often occurring in thin film solar cells.

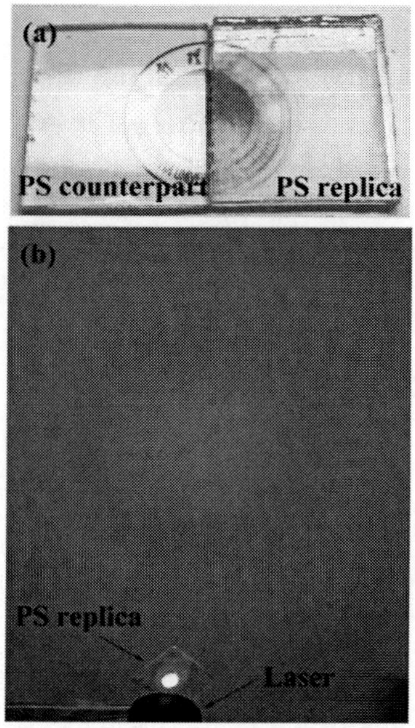

Figure 5. (a) Photograph of PS replica and PS counterpart exposed to white fluorescent light. (b) Diffraction phenomenon of PS replica.

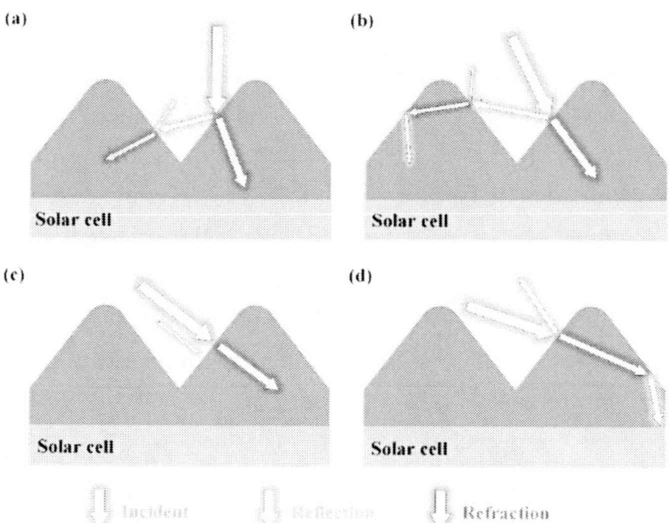

Figure 6. Schematics of light trapping effect of pyramids.

Table 3. Short-circuit current density J_{sc}, open-circuit voltage V_{oc}, fill factor FF, and power conversion efficiency PCE of pristine thin film solar cell and solar cell covered with PS replica and PS counterpart under normal incidence

Solar cell	J_{sc} (mA/cm^2)	V_{oc} (mV)	FF (%)	PCE (%)
Pristine	6.11	1.98	45.75	5.53
With PS counterpart	6.49	2.01	46.81	6.10
With PS replica	6.90	2.02	47.34	6.58

Enhanced Photocurrent and External Quantum Efficiency

The effect of the PS replica with pyramids as an antireflective protective layer on the photovoltaic performance of a thin film solar cell was investigated. Figure 7 shows the J-V and EQE curves of the thin film solar cell covered with the PS replica and PS counterpart, and Table 3 summarizes the main parameters obtained from J-V tests with normal incident. As shown in Figure 7(a) and Table 3, the pristine thin film solar cell exhibits an inferior photocurrent performance. The solar cell covered with the PS replica with the pyramids exhibits a short-circuit current density (J_{sc}), open-circuit voltage (V_{oc}), and fill factor (FF) of 6.90 mA/cm^2, 2.02 V, and 47.34%, respectively, resulting in a power conversion efficiency (PCE) of 6.58% (Table 3). In comparison to the solar cell covered with the PS counterpart, the J_{sc} of the solar cell covered with the PS replica is increased by 6.3% under normal incidence, leading to an about 7.9% enhancement of the PCE. The V_{oc} and FF do not show significant changes, which indicates that the improved PCE is mainly attributed to the increased light absorption at the active layer of the solar cell. The EQE curves depicted in Figure 7(b) show that the main portion of the photocurrent for the covered solar cell is generated in the wavelength range between 430 and 650 nm. The EQE is enhanced almost all over the utilized spectral range for the solar cell covered with the PS replica. This reveals a broadband enhancement starting at around 450 nm, which is in accordance with the light transmittance and reflectance spectra obtained for the PS replica and PS counterpart (Figure 4).

The superior performance of the solar cell covered with the PS replica can be ascribed to the light trapping effect via two means: the increased optical length and the total internal reflection of reflected or scattered light. The former suppresses the reflection at the interface of the protective layer (i.e., the PS replica) and air. More importantly, the latter can largely hinder the light reflected or scattered from the PS replica-solar cell interface from escaping to air, thus redirecting the light back to the solar cell.

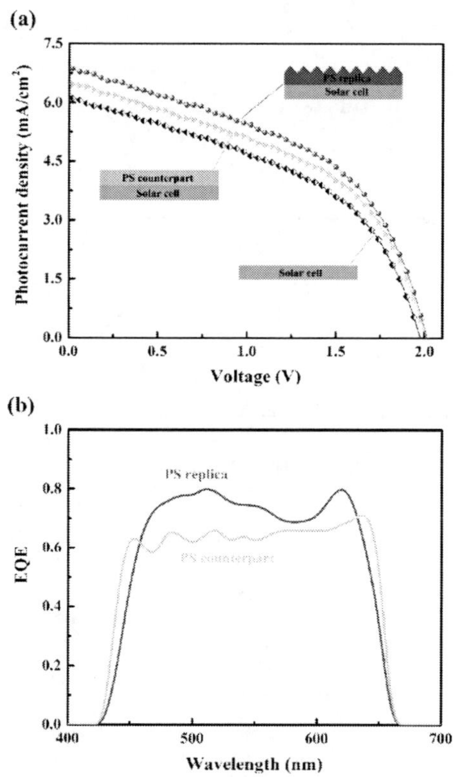

Figure 7. (a) J-V curves of pristine thin film solar cell and solar cell covered with PS replica and PS counterpart. (b) EQE curves of solar cell covered with PS counterpart and PS replica.

The J_{sc} and PCE of the solar cell covered with the PS replica and PS counterpart were measured at different incident angles (10°–70°) to investigate its incident angle-dependent photovoltaic performance. The results are shown in Figure 8, in which the results obtained at 0° are also given. As can be seen, the J_{sc} and PCE decrease simultaneously with increasing incident angle.

Obviously, the solar cell covered with the PS replica exhibits a superior photovoltaic performance compared to that covered with the PS counterpart over a broad range of incident angle. This enhanced performance can be attributed to the less angle-dependent transmittance characteristics of the PS replica (Figure 6). Therefore, the use of the PS replica can enhance the photovoltaic performance of the solar cell for an entire day and the seasons when incident angle varies. Furthermore, the high-transmittance polymeric film with micropyramids can be applied to other various solar cell systems as well.

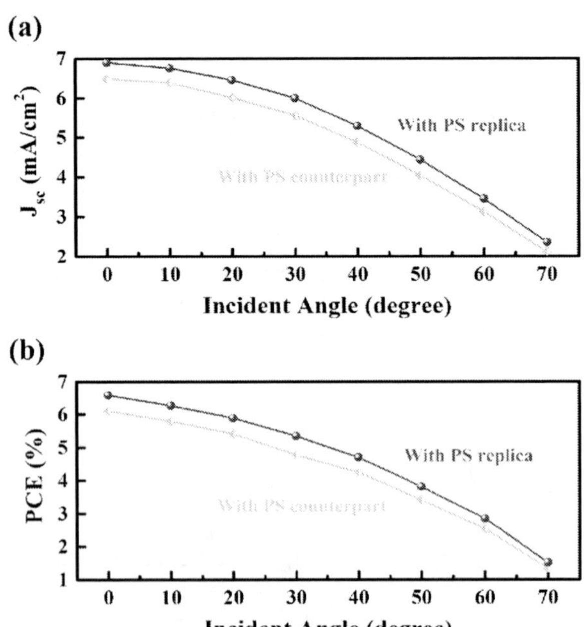

Figure 8. Angle-dependent (a) J_{sc} and (b) PCE of thin film solar cell covered with PS replica and PS counterpart.

CONCLUSION

The pyramids on the etched silicon wafer surface have the hemlines of about 1–7 μm and the heights of about 1–5 μm and play a crucial role in suppressing reflection of light. The staggered and densely arranged negative pyramids on the prepared nickel mold insert surface are the inverse pattern of the pyramids on the etched silicon wafer. Applying μ-ICM, the pyramids on the nickel mold insert are accurately and efficiently transcribed to the PS replica surface. The PS replica surface exhibit similar pyramids with the etched silicon wafer surface, the trapezoidal surfaces and arc-shaped tops of which are much smooth.

The PS replica surface exhibits a reflectance of about 5% and an increased transmittance of 93% in the wavelength range of 400–900 nm. This enhanced light trapping effect can be ascribed to the staggered and densely arranged, differently-sized pyramids. Especially, the thin film solar cell covered with the PS replica exhibits increased power conversion efficiency enhancement of up to 7.9% and enhanced EQE under normal incidence compared to that covered with the PS counterpart. The solar cell covered with the PS replica also shows an excellent photovoltaic performance over a wide range of incident angles (0°–70°). The proposed strategy provides a fast, high-efficient, low-cost, and large-scale route for fabricating antireflective protective layers in a single step.

ACKNOWLEDGMENTS

Financial support provided by the National Natural Science Foundation of China (contract grant number: 51533003) and Guangdong Provincial Natural Science Foundation (contract grant number: 2016A030308018) is gratefully acknowledged.

REFERENCES

[1] Llopis, F., Tobias, I. Texture profile and aspect ratio influence on the front reflectance of solar cells, *J. Appl. Phys.*, 100 (2006) 124504.

[2] Li, Y. F., Zhang, J. H., Yang, B. Antireflective surfaces based on biomimetic nanopillared arrays, *Nano Today*, 5 (2010) 117–127.

[3] Hayoung, P., Soonwoo, K., Joon Sung, L., Hee Jin, L., Sewang, Y., Donghwan, K. Improvement on surface texturing of single crystalline silicon for solar cells by saw-damage etching using an acidic solution, *Sol. Energ. Mat. Sol. C.,* 93 (2009) 1773–1778.

[4] Zhao, X., Sanchez, B. M., Dobson, P. J., Grant, P. S. The role of nanomaterials in redox-based supercapacitors for next generation energy storage devices, *Nanoscale,* 3 (2011) 839–855.

[5] Brodoceanu, D., Alhmoud, H. Z., Elnathan, R., Delalat, B., Voelcker, N. H., Kraus, T. Fabrication of silicon nanowire arrays by near-field laser ablation and metal-assisted chemical etching, *Nanotechnology,* 27 (2016) 075301.

[6] Toma, M., Loget, G., Corn, R. M. Flexible teflon nanocone array surfaces with tunable superhydrophobicity for self-cleaning and aqueous droplet patterning, *ACS Appl. Mater. Inter.*, 6 (2014) 11110–11117.

[7] Munoz, D., Carreras, P., Escarre, J., Ibarz, D., de Nicolas, S. M., Voz, C., Asensi, J. M., Bertomeu, J. Optimization of KOH etching process to obtain textured substrates suitable for heterojunction solar cells fabricated by HWCVD, *Thin Solid Films*, 517 (2009) 3578–3580.

[8] Chu, K., Wang, J. S., Tsai, Z. Y., Lee, C. K. A simple and cost-effective approach for fabricating pyramids on crystalline silicon wafers, *Sol. Energ. Mat. Sol. C.*, 93 (2009) 1276–1280.

[9] Hua, X. S., Zhang, Y. J., Wang, H. W. The effect of texture unit shape on silicon surface on the absorption properties, *Sol. Energ. Mat. Sol. C.,* 94 (2010) 258–262.

[10] Wu, D. M., Sun, J. Y., Liu, Y., Yang, Z. Z., Xu, H., Zheng, X. T., Gou, P. S. Rapid fabrication of microstructure on PMMA substrate

by the plate to plate transition-spanning isothermal hot embossing method nearby glass transition temperature, *Polym. Eng. Sci.*, 57 (2017) 268–274.

[11] Peng, K., Fang, H., Hu, J., Wu, Y., Zhu, J., Yan, Y., Lee, S. Metal-particle-induced, highly localized site-specific etching of Si and formation of single-crystalline Si nanowires in aqueous fluoride solution, *Chem. Eur.* J., 12 (2006) 7942–7947.

[12] Srivastava, S. K., Kumar, D., Singh, P. K., Kar, M., Kumar, V., Husain, M. Excellent antireflection properties of vertical silicon nanowire arrays, *Sol. Energ. Mat. Sol. C.*, 94 (2010) 1506–1511.

[13] Chen, Y. Q., Elshobaki, M., Ye, Z., Park, J. M., Noack, M. A., Ho, K. M., Chaudhary, S. Microlens array induced light absorption enhancement in polymer solar cells, *Phys. Chem. Chem. Phys.*, 15 (2013) 4297–4302.

[14] Zhou, W. C., Raasch, T. W., Yi, A. Y. Design, fabrication, and testing of a Shack-Hartmann sensor with an automatic registration feature, *Appl. Opt.,* 55 (2016) 7892–7899.

[15] Campbell, P., Green, M. A. Light trapping properties of pyramidally textured surfaces, *J. Appl. Phys.*, 62 (1987) 243–249.

[16] Green, M. A., Jianhua, Z., Wang, A., Wenham, S. R. Very high efficiency silicon solar cells-science and technology, *IEEE Trans. Electron Devices,* 46 (1999) 1940–1947.

[17] Dewan, R., Vasilev, I., Jovanov, V., Knipp, D. Optical enhancement and losses of pyramid textured thin-film silicon solar cells, *J. Appl. Phys.*, 110 (2011) 013101.

[18] Kapur, J., Proost, K., Smith, C. A. Determination of moisture ingress through various encapsulants in glass/glass laminates, in: *Proc. 34th IEEE Photovoltaic Specialists Conf.*, 2009, pp. 001210–001214.

[19] Hiller, J., Mendelsohn, J. D., Rubner, M. F. Reversibly erasable nanoporous anti-reflection coatings from polyelectrolyte multilayers, Nature Mater., 1 (2002) 59–63.

[20] Pintossi, D., Iannaccone, G., Colombo, A., Bella, F., Valimaki, M., Vaisanen, K. L., Hast, J., Levi, M., Gerbaldi, C., Dragonetti, C., Turri, S., Griffini, G. Luminescent downshifting by photo-induced

sol-gel hybrid coatings: accessing multifunctionality on flexible organic photovoltaics via ambient temperature material processing, *Adv. Electron. Mater.,* 2 (2016) 1600288.

[21] Bella, F., Lamberti, A., Bianco, S., Tresso, E., Gerbaldi, C., Pirri, C. F. Floating, flexible dye-sensitized solar-cell architecture: the way of near-future photovoltaics, *Adv. Mater. Technol.,* 1 (2016) 1600002.

[22] Lim, J. H., Leem, J. W., Yu, J. S. Solar power generation enhancement of dye-sensitized solar cells using hydrophobic and antireflective polymers with nanoholes, *RSC Advances*, 5 (2015) 61284–61289.

[23] Leem, J. W., Choi, M., Yu, J. S. Multifunctional microstructured polymer films for boosting solar power generation of silicon-based photovoltaic modules, *ACS Appl. Mater. Inter.*, 7 (2015) 2349–2358.

[24] Lee, S. H., Han, K. S., Shin, J. H., Hwang, S. Y., Lee, H. Fabrication of highly transparent self-cleaning protection films for photovoltaic systems, *Prog. Photovoltaics,* 21 (2013) 1056–1062.

[25] Kang, S. M., Ahn, N., Lee, J. W., Choi, M., Park, N. G. Water-repellent perovskite solar cell, *J. Mater. Chem.* A, 2 (2014) 20017–20021.

[26] Biswas, Peer, R., Park, J. M., Shinar, R., Shinar, J. Light management in perovskite solar cells and organic LEDs with microlens arrays, *Opt. Express,* 25 (2017) 10704–10709.

[27] Bella, F., Pugliese, D., Zolin, L., Gerbaldi, C. Paper-based quasi-solid dye-sensitized solar cells, *Electrochim. Acta*, 237 (2017) 87–93.

[28] Farhana, N. K., Khanmirzaei, M. H., Ramesh, S., Ramesh, K. Exploration on polypropylene carbonate polymer for gel polymer electrolyte preparation and dye-sensitized solar cell application, *J. Appl. Polym. Sci.*, 134 (2017) 45091.

[29] Hiroshima, H., Komuro, M. Control of bubble defects in UV nanoimprint, *Jpn. J. Appl. Phys., Part 1*, 46 (2007) 6391–6394.

[30] Liu, H. B., Gong, H. Q., Ramalingam, N., Jiang, Y., Dai, C. C., Hui, K. M. Micro air bubble formation and its control during polymerase chain reaction (PCR) in polydimethylsiloxane (PDMS) microreactors, *J. Micromech. Microeng.*, 17 (2007) 2055–2064.

[31] Andersen, N. K., Taboryski, R. Multi-height structures in injection molded polymer, *Microelectron. Eng.*, 141 (2015) 211–214.

[32] Guan, W. S., Huang, H. X., Chen, A. F. Tuning 3D topography on biomimetic surface for efficient self-cleaning and microfluidic manipulation, *J. Micromech. Microeng.,* 25 (2015) 035001.

[33] Chen, F., Huang, H. X. Rapid fabrication of T-shaped micropillars on polypropylene surfaces with robust Cassie-Baxter state for quantitative droplet collection, *J. Phys. Chem. C,* 120 (2016) 1556–1561.

[34] Saarikoski, I., Suvanto, M., Pakkanen, T. A. Modification of polycarbonate surface properties by nano-, micro-, and hierarchical micro-nanostructuring, *Appl. Surf. Sci.*, 255 (2009) 9000–9005.

[35] Fuh, Y. K., Peng, C. C., Huang, C. T. Rapid replication and facile modulation of subwavelength antireflective polymer film using injection nanomolding and optical property of multilayer coatings, *Nanoscale Res. Lett.,* 8 (2013) 407.

[36] Guan, W. S., Huang, H. X., Wang, B. Poiseuille/squeeze flow-induced crystallization in microinjection-compression molded isotactic polypropylene, *J. Polym. Sci., Part B: Polym. Phys.,* 51 (2013) 358–367.

[37] Deinega, A., Valuev, I., Potapkin, B., Lozovik, Y. Minimizing light reflection from dielectric textured surfaces, *J. Opt. Soc. Am. A Opt. Image Sci. Vis.,* 28 (2011) 770–777.

In: Polymer Surfaces
Editor: Valente Erasmo Silva
ISBN: 978-1-53614-309-6

Chapter 4

Gradient Wetting State on Cicada Wing Surface and Rapid Fabrication of Its Nanostructure with Hydrophobicity and Antireflectivity on Polystyrene Surface

Han-Xiong Huang*[*] *and Heng Xie
Lab for Micro/Nano Molding and Polymer Rheology,
The Key Laboratory of Polymer Processing Engineering
of the Ministry of Education, South China University of Technology,
Guangzhou, People's Republic of China

Abstract

As a result of natural selection, cicadas can easily remove water droplets and dust from their wing surfaces. The fine nanostructure on the cicada wing, which exhibits hydrophobicity and antireflectivity, is carefully examined in this work. A gradient wetting state along the wing veins is revealed for the first time. That is, the Wenzel state (hydrophilic)

[*] Corresponding author: Email: mmhuang@scut.edu.cn.

is transitioned to the Cassie-Baxter state (superhydrophobic) from foot to apex of the wing. Superhydrophobic region exhibits a lower adhesion force to the droplets than hydrophilic region. So the accumulated droplets easily roll off from superhydrophobic to hydrophilic regions, and the light weight state on the wing can be maintained under its slight shaking.

Further, a promising strategy is proposed for facilely and successively replicating the natural functional nanostructure of the cicada wing onto polystyrene (PS) surfaces. First, a nickel replica with tapered nanopores is fabricated by combining electroless plating and subsequent electroplating with the natural cicada wing as an original template. Then, using microinjection compression molding, with the nickel replica as a template, the tapered nanopores are transcribed onto the PS surface, resulting in orderly and densely arranged nanopillars with a mean diameter of about 156 nm and a mean pitch of about 180 nm. Interestingly, the PS replica surface exhibits a water contact angle of 143° ± 2° and a reflectance of about 4% in a wavelength range of 400–1000 nm. These results mean that the bionic PS replica not only inherits the nanostructure of the natural wing, but also its hydrophobic and antireflective properties. The mechanisms for the hydrophobic and antireflective properties are revealed via composite wetting interface and effective medium layer on the replica surface, respectively.

This work may direct the design of gradient wetting surfaces by mimicking the nanopillar structure on cicada wing. The proposed fast and efficient replication strategy can be an excellent candidate for mimicking bio-inspired functional micro/nanostructures without complicated procedures and expensive materials.

Keywords: cicada wing, gradient wetting state, nanostructured surface, biomimetic, hydrophobicity, antireflection, microinjection compression molding

INTRODUCTION

In nature, numerous of creatures possess various excellent surface characteristics, resulting from the presence of almost perfect micro/nanostructures over a wide range [1]. It stands to reason that the perfect surface is an ideal innovation source for novel materials, structures, methods, and tools. Examples include continuous water transport on the peristome surface of *Nepenthes alata* [2], structure-based color in butterfly wing scales [3], antifogging property of mosquito compound eyes [4], water-repellent legs of water striders [5], antireflective property of moth-eye and glasswing butterfly [6, 7], self-cleaning of lotus leaves [8], and water pinning property on rose petals [9]. Researches on how to replicate these structures into new functional surfaces are coming to the forefront of research in the micro/nano manufacturing technologies. The artificial micro/nanostructures have an extensive application prospect, such as observing systems [10], solar cells [11], optical fibers [12], spectral conversion [13], and droplet transport [14]. Great attentions have been paid to mimic micro/nanostructures inspired by biological prototypes found in nature during the last few decades, such as compound eyes [15, 16], antireflective hierarchical arrays [17], photonic crystal structures in butterfly wing scales [18], bionic nanopillar arrays [19], and honeycomb structured films [20].

As one of typical biological prototypes in the field of bionics, cicada wings possess remarkable properties, such as optical trapping, hydrophobicity, and bacteria killing [21–23]. By virtue of an impedance matching mechanism, cicada has a camouflage ability associated with the antireflective function of tapered nanostructure (TNS) on the wings to avoid itself to be captured. The TNS on the wings can be approximated as a set of multiple layers with a gradient refractive index between substrate and ambient [24]. In view of gradient refractive index of the multiple layers, light tends to bend progressively and reflection can be suppressed to a very low level when it hits the wing surface. Moreover, cicadas inevitably contend with rain, fog, and so on. These environmental surroundings cause problems for cicadas, both during flight and at rest. Fortunately, cicada wings still maintain a clean and light weight state in nature. Although the wetting behaviors were investigated for the cicada family [23, 25], mechanism for maintaining the light weight state on the wing surface is still not clearly understood. In this work, the wetting state transition along the wing veins is revealed for the first time. The mechanisms for transporting droplets and maintaining the light weight state on the wing surface are analyzed.

Zhang and coworkers [26] used natural cicada wing as a bio-template and prepared Au film with cicada wing inspired nanostructures on film surfaces. Functional surfaces with cicada wing inspired tapered micro/nanostructures have been fabricated using several methods, e.g., hot embossing [27], soft-lithography [28], self-assembly [29], self-masked dry etching [30], and electrodeposition [31]. The available works mainly focus on the antireflective properties of the fabricated tapered micro/nanostructures. The mechanisms for the hydrophobicity and antireflectivity on cicada wings and fabricated surfaces with cicada wing inspired nanostructure, unfortunately, are still unclear. Moreover, either special equipment or complex control is required in the bionic replication processes, which limits the applications of bio-inspired micro/nanostructured surfaces with excellent functions. Therefore, successive and facile methods need to be proposed for replicating the subtle nanostructures from a bionic template.

To pursue facile fabrication of the nanostructure with hydrophobicity and antireflectivity on a cicada wing, a strategy, which combines electroless plating, electroplating, and microinjection compression molding (μ-ICM), is proposed in this work. μ-ICM is typically used in high-efficient, high-precise, low-cost, and large-scale production and quite qualified for rapid fabrication. Using the proposed strategy, polystyrene (PS) replicas with orderly and densely arranged nanopillars on their surfaces are prepared, which exhibit hydrophobic and antireflective properties that inherit from the cicada wings.

EXPERIMENTAL

The wings of cicada *Cryptotympana atrata fabricius* were taken as original wings. The natural wings were cleaned cautiously. First, the wings were washed by ultrasonic oscillation for 20 min with deionized water to remove contaminants absorbed physically on their surfaces. They were then ultrasonically cleaned in acetone for 20 min to remove the stains, proteins, and fat that were stuck to the nanopillars on the wings. Finally, they were ultrasonically cleaned in the deionized water again for 10 min to remove residual acetone, followed by drying in a vacuum oven at 40°C.

Process for Fabricating Bio-Inspired Polystyrene Replica

The overall process for fabricating the bio-inspired PS replica with nanopillars replicating from the cicada wing consists of three steps: natural wing template pretreatment, nickel replica preparation (Figure 1(1)), and PS replica molding (Figure 1(2)).

The aforementioned cleaned natural wings were treated by sensitization and activation, in which the compositions and concentrations are listed in Table 1. Specifically, the wings were first immersed in the sensitization solution for 10 min at 25°C and then in the activation solution for 5 min at 25°C. In the sensitization process, sodium dodecyl sulfonate ($C_{12}H_{25}SO_4Na$) was used as a surfactant to improve hydrophilicity of the wing surfaces. So the sensitization and activation solutions could fully penetrate into the gaps between the nanopillars on the wings, thus ensuring the nanopillars to be entirely covered by palladium chloride ($PdCl_2 \cdot H_2O$) in the activation solution. Then the natural wings were dried naturally in air. The covered palladium catalytic nuclei on the wing can be used as release of metal ions in the reduction reaction of subsequent electroless plating.

The nickel replica was fabricated via combining the electroless plating with electroplating, as shown in Figure 1(1). The compositions and concentrations in the used solutions are listed in Table 2. The aforementioned pretreated natural wing was fixed on a supporting glass-slide and then immersed into the electroless nickel plating solution with magnetic stirring at 150 rpm under 40°C. The solution was adjusted to a pH value of 8.5 using ammonia water. Nickel sulfate ($NiSO_4 \cdot 6H_2O$) and sodium hypophosphite ($NaH_2PO_2 \cdot H_2O$) were used as main salt and reductant, respectively, in the electroless plating solution. After 10 min of electroless plating, the wing was taken out from the solution and rinsed thoroughly with deionized water. A conductive nickel layer was deposited on the natural wing surface through the above electroless plating (Figure 1(b)). The electroplating was subsequently performed with the conductive nickel covered wing as a cathode and a Ni-S plate as an anode at 50°C. A power supply (IT7000 DC Power Supply, ITECH Electronic Co. Ltd,

USA) was used to provide a constant current density of 4 A/dm^2 for the whole electroplating process. To guarantee the uniformity of the electrolyte solution, a magnetic stirring at 150 rpm was applied. After 120 min of the electroplating, a nickel layer (Figure 1(c)) with an approximately 400 μm thickness was deposited on the conductive nickel layer in the electroplating solution. After cleaning and drying, the nickel layer was peeled off the wing. At this point, a nickel replica (Figure 1(d)) was fabricated.

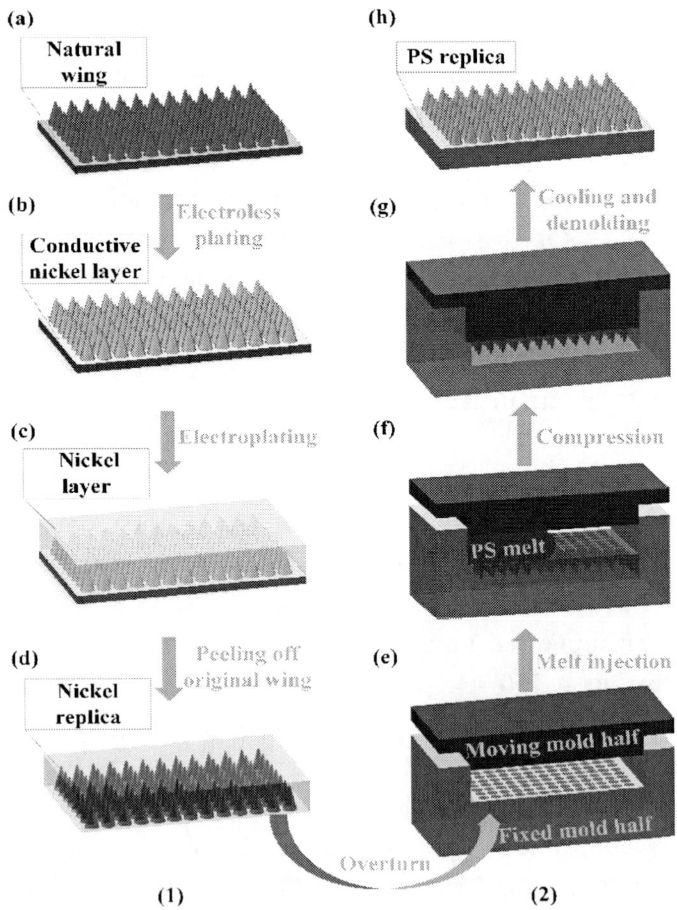

Figure 1. Schematics of fabrication process for bio-inspired PS replica. (a) 3D model of natural cicada wing, (b) conductive layer deposition, (c) nickel deposition on conductive layer surface, (d) nickel replica, (e, f, g) μ-ICM process and (h) final PS replica.

Table 1. Compositions and concentrations in sensitization and activation solutions for wing retreatment

Step	Composition	Concentration
Sensitization	$SnCl_2 \cdot H_2O$	10 g/L
	HCl	40 mL/L
	$C_{12}H_{25}SO_4Na$	1 g/L
Activation	$PdCl_2 \cdot H_2O$	1 g/L
	HCl	10 mL/L

Table 2. Compositions and concentrations in electroless plating and electroplating solutions for nickel replica fabrication

Step	Composition	Concentration
Electroless nickel plating	$NiSO_4 \cdot 6H_2O$	30 g/L
	$NaH_2PO_2 \cdot H_2O$	30 g/L
	$Na_3C_6H_5O_7 \cdot 2H_2O$	20 g/L
Electroplating	$Ni(SO_3NH_2)_2 \cdot 4H_2O$	300 g/L
	$NiCl_2 \cdot 6H_2O$	15 g/L
	H_3BO_3	20 g/L

The bionic PS replicas were molded via the μ-ICM technology, as shown in Figure 1(2). PS (grade N1841H, Hong Kong Petrochemical Co. Ltd) with a refractive index of 1.6 was used as the material for molding the replicas. The equipment used in the μ-ICM process comprised an 80 ton injection molding machine (KM80SP180CX, Krauss-Maffei, Germany) and a microinjection compression mold equipped with a temperature control apparatus. As a template, the previously prepared nickel replica was mounted on the mold cavity surface (Figure 1(e)). In the μ-ICM process, the mold cavity was first partially filled with PS melt when the mold was partially closed (Figure 1(f)). Then the mold was compressed in the thickness direction to finish melt filling (Figure 1(g)). Finally, the melt was packed under the compression force and simultaneously cooled down. The molding parameters, including melt temperature (240°C), injection rate (154 cm^3/s), compression force (280 kN), and mold temperature (100°C), were kept constant. It should be highlighted here that the molding cycle of each replica was about 50 s. For comparison, PS counterparts with a smooth surface were molded per the same protocol without the nickel replica as template.

Characterization

The nanostructures on the natural cicada wings and the prepared nickel replicas and PS replicas were characterized using a scanning electronic microscopy (SEM; Nova NanoSEM 430, FEI Co., Netherlands) and atomic force microscopy (AFM; Multimode 8, Bruker Corporation, Germany). Prior to the SEM observation, the samples were coated with gold by ion sputtering. Quantitative analyses of the nanostructures on the used wings and the prepared nickel replica and PS replica surfaces were conducted by using the Nanoscope analysis software (Version 1.6, Bruker Corporation, Germany).

The contact angles (CAs) and roll off angles (RAs) of droplets on the wings and the PS replica surfaces were measured using an automatic CA testing apparatus (OCA 40, Data Physics Corporation, Germany). Both the CA and RA measurements were conducted at three different locations on the wings and the replica surfaces to obtain the average values.

The fog condensation and droplet transporting behaviors on the wings were tested (as displayed in Figure 2). An ultrasonic humidifier was employed to mimic a fog flow composed of numerous tiny water droplets with diameters less than 10 μm. It takes 60 ± 5 min to transform 280 mL of water to fog. The wing was held vertically and kept 6 cm away from the exit hole of the mist pipe in the humidifier. A high-speed camera (TS5, Fastec Imaging Corporation, USA) was used to record the condensation and transportation processes of the water droplets during testing.

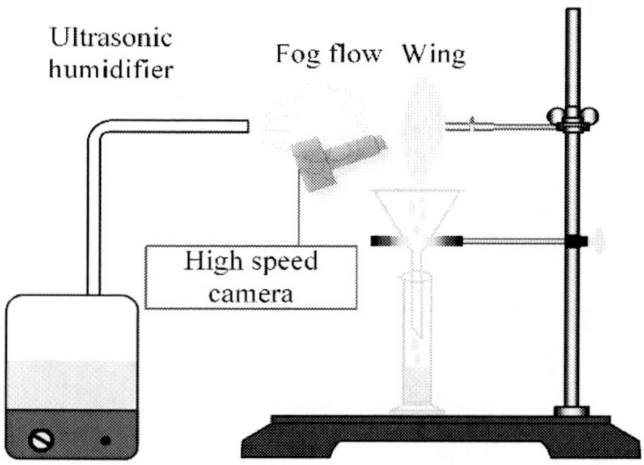

Figure 2. Schematics for testing fog condensation and droplet transporting behaviors on cicada wings.

The element compositions of the nickel replicas were analyzed using an energy dispersive spectroscopy (EDS; Inca Penta FET x3, Oxford Instruments, UK) coupled with an SEM.

The reflectance spectrum measurements for the wing and replica surfaces were performed using an UV/VIS/NIR spectrometer (UV-3150, Shimadzu Corporation, Japan) equipped with an integrating sphere attachment in a wavelength range of 400–1000 nm.

Results and Discussion

Gradient Wetting State and Mechanisms for Transporting Droplets and Maintaining Light Weight State on Cicada Wings

Four typical regions were selected along the wing veins, the spread of which is about 60 mm. Figure 3 shows the wetting states for water droplets of about 8–10 μL, CAs for 4 μL water droplet, and nanostructures at the four regions. As can be clearly seen from Figure 3(a) and (b), the droplets exhibit a hydrophilic state at regions 1 and 2, whereas an obvious hydrophobic state at regions 3 and 4. The nanopillars orderly arrange on the wing surface (Figure 3(c)). The water droplets at regions 1 and 2 are easy to penetrate into the bottom of the nanopillars mainly due to their large gaps (Figure 3(c)), thus forming the Wenzel state (Figure 3(d)). The nanopillars with small gaps (Figure 3(c)) at regions 3 and 4 can effectively hinder the droplets from wetting the lower parts of the inter-nanopillar areas where air pockets are formed. This reduces the solid-liquid contact area and results in the formation of the Cassie-Baxter state (Figure 3(d)). Therefore, a gradient wetting behavior appears along the wing veins.

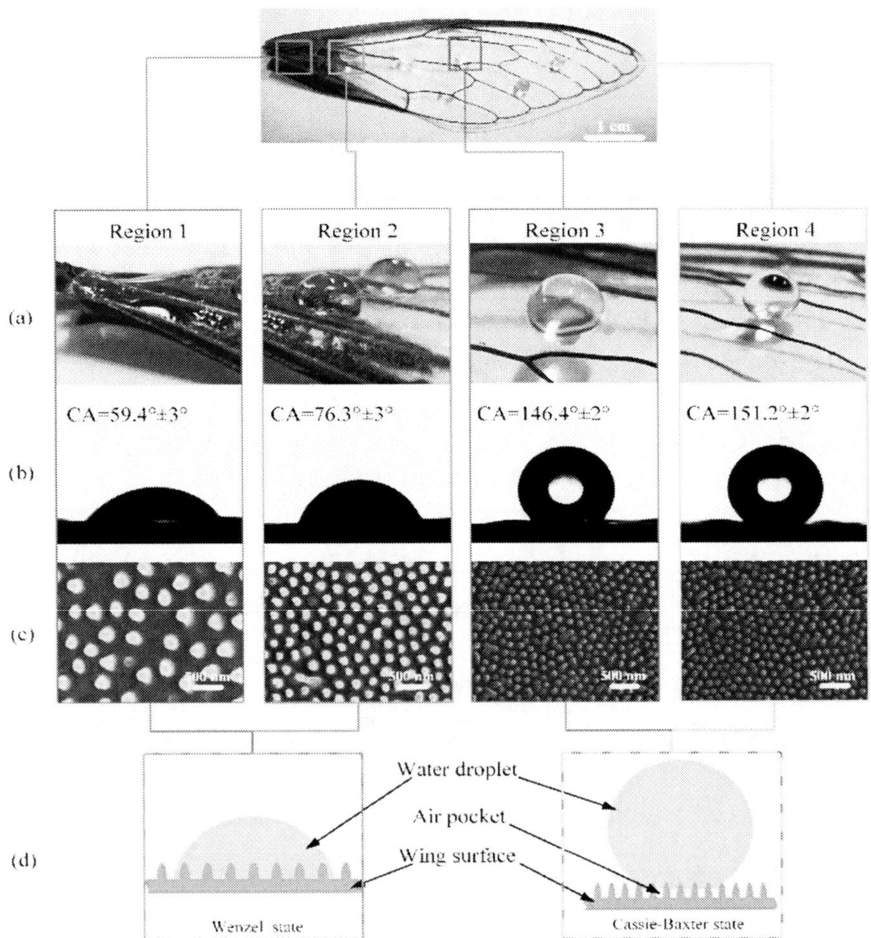

Figure 3. (a) Wetting states, (b) contact angles, (c) nanostructures at typical four regions along cicada wing veins and (d) schematics of wetting states for (1) regions 1and 2 (Wenzel) and (2) regions 3 and 4 (Cassie-Baxter).

The fog condensation and droplet transporting behaviors on the wing surfaces were recorded via focusing on the transitional area between regions 2 and 3. Typical snapshots taken from recorded videos are shown in Figure 4(a) and (b), respectively. Obviously, tiny water droplets begin to condense on the wing surface (0.5 min) under the fog flow and then the adjacent tiny droplets coalesce with each other to form a larger droplet (2 min), as shown in Figure 4(a).

When the droplet grows large enough, it rolls down from the top (regions 3 and 4) to end (regions 1 and 2) of the wing due to the gravitation effect. Path I (Figure 4(b)) was chosen to depict the droplet transportation process on the wing surface. Droplet 1 merges with droplet 2 along path I. Then the accumulated droplet rolls down and further merges with droplet 3, thus substantially increasing the volume of the droplets at regions 1 and 2. Finally, the accumulated droplets gain increased transportation rate and sweep the wing surface along path I, resulting in a remarkable fog collection efficiency.

Based on the aforementioned observed phenomenon, the mechanism for transporting droplets and maintaining light weight state on the wing can be schematically depicted in Figure 5. Tiny droplets condense and merge under the fog flow on the wing surface (Figure 5(a) and (b)). Water droplets penetrate into the bottom of the nanopillars at regions 1 and 2, and only penetrate into the upper portion of the nanopillars at regions 3 and 4. That is, regions 3 and 4 exhibit a lower adhesion force (F_a) to the droplets than regions 1 and 2.

When the accumulated droplets at regions 4 and 3 reach the corresponding critical volumes, they easily roll to regions 2 and 1. By swallowing the droplets from regions 3 and 4, the water droplets at regions 1 and 2 increase in their volumes, and so easily roll off under a slight shaking of the wing. Thus, dust and pollen can be easily washed away by the rolling water droplets and the wing can be maintained in the light weight state (Figure 5(d)).

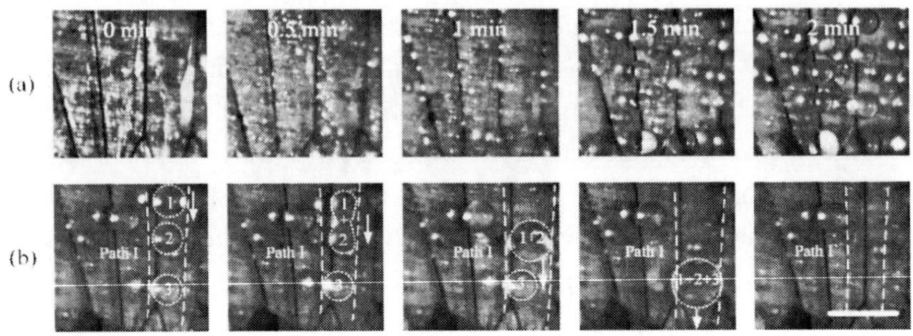

Figure 4. Snapshots of (a) fog condensation process and (b) droplets transport process along pass I on cicada wing (Scale bar: 5 mm).

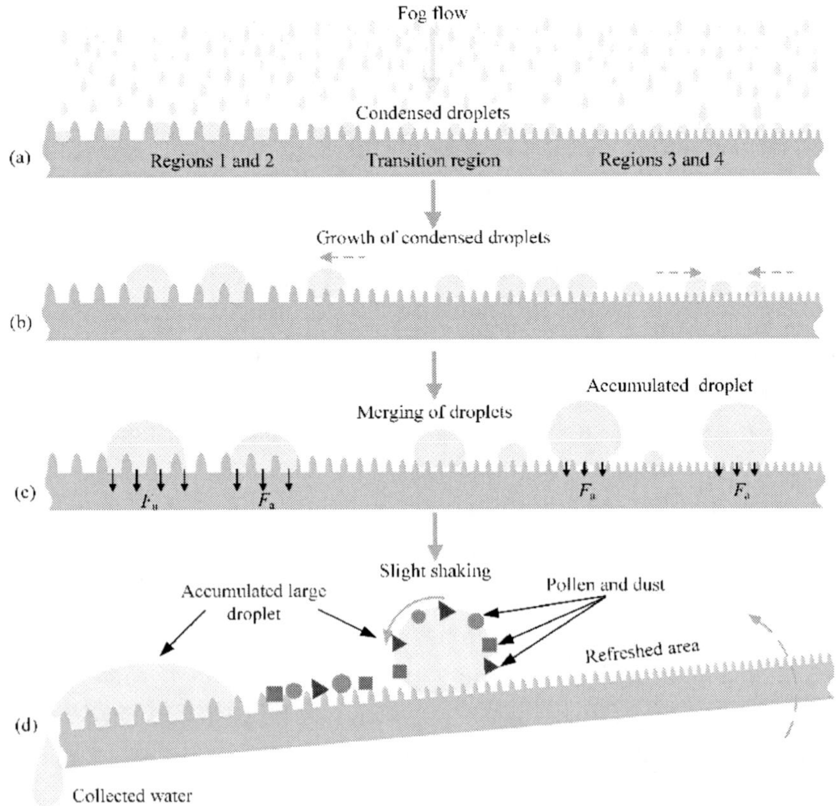

Figure 5. Schematics for transporting droplets and maintaining light weight state on cicada wing.

Nanostructure, Wettability and Antireflectivity on Cicada Wings

Further, region 3 on the cicada wing was carefully observed to reveal the correlation between its properties and nanostructure. Figure 6(a) shows the digital picture of the used cicada that died of natural causes. As can be clearly seen from Figure 6(b) and (c), the nanopillars, which have a mean diameter of about 160 nm and a mean pitch of about 180 nm, exhibit tapered structure, namely the TNS.

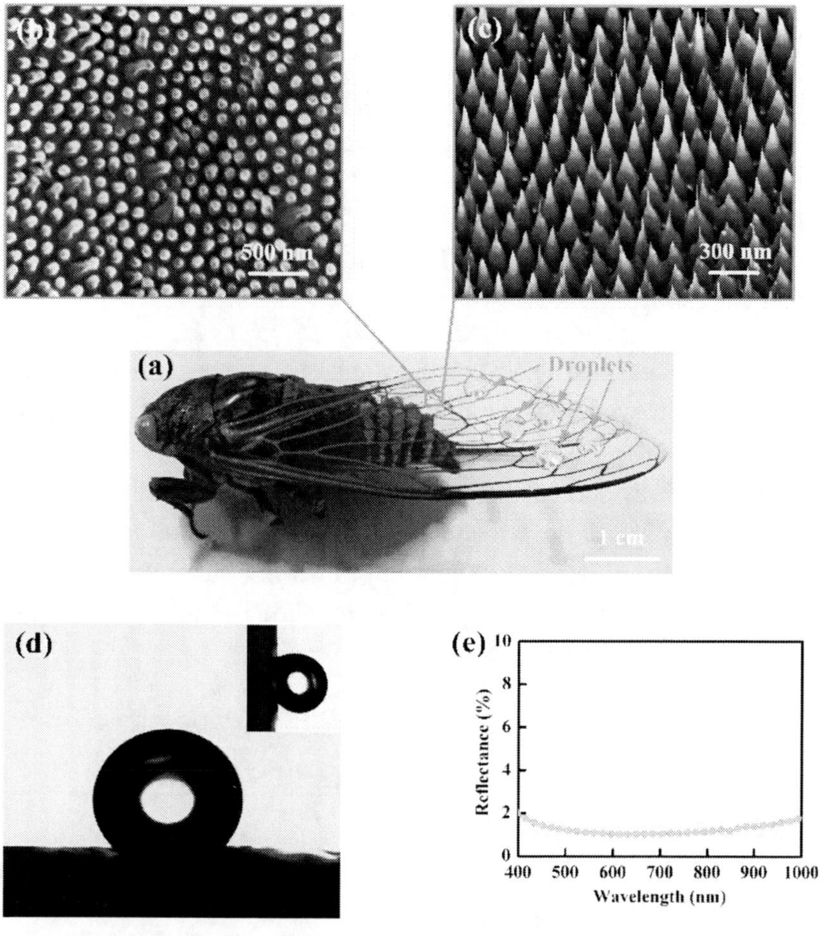

Figure 6. (a) Digital picture of natural cicada; (b) SEM and (c) AFM images, (d) wetting state of 4 μL droplet, and (e) reflectance spectrum on its wing.

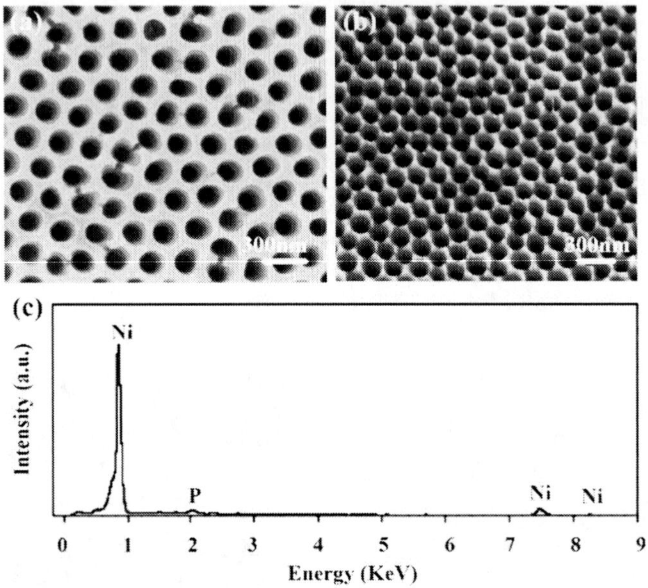

Figure 7. (a) SEM and (b) AFM images of nickel replica, and (c) its energy dispersive spectroscopy spectrum.

The aforementioned TNS on the cicada wing surface is closely correlated with its wettability and antireflectivity. A 4 μL droplet dropped at the selected region on the wing surface exhibits a CA of 145° ± 2° and an RA of > 90°, as shown in Figure 6(d). The reflectance of the cicada wing is less than 2% in the wavelength range of 400–1000 nm that covers the visible wavelength (400–700 nm), as shown in Figure 6(e). Apparently, this antireflective property helps the cicada to hide from its natural enemies in the visible wavelength range.

Nanostructures of Fabricated Nickel and PS Replicas

The nickel replicas were fabricated with region 3 on the cicada wing as an original template. Figure 7(a) and (b) shows the typical SEM and AFM images of the fabricated nickel replica. It can be found that tapered nanopores with a mean diameter of about 165 nm and a mean pitch of about 180 nm orderly distribute in dense rows throughout the nickel replica. That is, the nanostructure on the nickel replica is consistent with that on the natural wing surface shown in Figure 6(b) and (c). This indicates that the TNS on the natural wing is accurately replicated on the nickel replica with an inverse pattern via the electroless plating and electroplating. Moreover, by means of atom by atom depositing the nickel layer on the natural wing, the prepared nickel replica exhibits a smooth surface and sufficient strength. The EDS spectrum of the nickel replica (Figure 7(c)) indicates that its main component is nickel (98.53%) and it also contains small amounts (1.47%) of phosphorus. The cicada wing is composed of chitin [22], whereas no carbon peak appears at the EDS spectrum of the nickel replica, which demonstrates that the nickel replica does not contain natural wing residue and has high purity.

The testing demonstrates that the nickel replica has a tensile strength of about 215 MPa, which means that it can be repeatedly used. The nickel replica after ten cycles of molding was investigated for the further analysis of its durability. As can be seen in Figure 8, no carbon peak appears at its EDS spectrum, which demonstrates that no PS residue exists on the repeatedly used nickel replica. Interestingly, oxidation occurs on the nickel replica surface during μ-ICM. The SEM image shown in Figure 8 clearly demonstrates that the nanopores on the repeatedly used nickel replica still remain intact. Therefore, the nickel replica prepared herein is reusable.

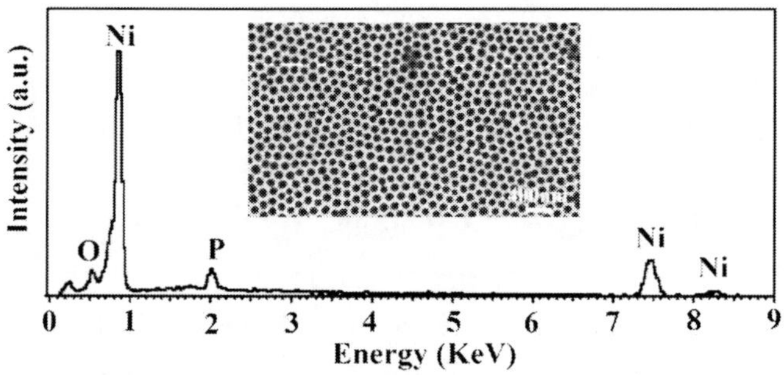

Figure 8. Energy dispersive spectroscopy spectrum and (inset) SEM image of nickel template after ten cycles molding.

The tapered nanopores in the nickel replica, which are inherited from the natural wing, are filled by the PS melt at a sufficient compression force in the μ-ICM process. Therefore, the nanopillars on the final PS replica are the inverse of the nanopores in the nickel replica. Figure 9(a) is the SEM image of the PS replica at low magnification, which clearly indicates that the TNS on the natural wing is successfully transferred to the PS replica using μ-ICM with the nickel replica as the template. As can be seen from Figure 9(b) and (c), the nanopillars on the PS replica possess a typical mean diameter of about 156 nm and a mean pitch of about 180 nm. To further characterize the inherited accuracy, the nanopillar heights on both natural wing and final PS replica surfaces were gathered from the AFM images (Figure 6(c) and 9(c)), and the results are shown in Figure 10. It is clear that the height histograms for the nanopillars on the PS replica and natural wing exhibit a similar profile, which demonstrates that the PS replica inherits the nanostructure on the natural wing with remarkable accuracy. From the above results, it can be observed that the nanopillars on the PS replica surface are consistent with those on the natural wing shown in Figure 6(b) and (c). Therefore, the nanostructure on the natural wing surface is replicated on the PS replica surface with a high fidelity by μ-ICM.

Hydrophobicity and Antireflectivity of PS Replica and Relevant Mechanism Analyses

A 4-μL droplet dropped on the smooth surface of the PS counterpart exhibits a CA of 83° ± 2° and an RA of >90° (Figure 11(a)). The nanostructured PS replica surface exhibits a CA of 143° ± 2° and an RA of >90° (Figure 11(b)), which is very similar to those exhibiting on the natural wing surface shown in Figure 6(d). That is, the hydrophilicity on the PS counterpart surface is transformed to the hydrophobicity on the PS replica surface, which is apparently ascribed to the bionic TNS on the PS replica. The reflectance spectra of the PS counterpart and PS replica surfaces are shown in Figure 11(c). The average reflectances of the PS counterpart and bionic PS replica are about 9% and 4%, respectively, in the wavelength range of 400–1000 nm, which demonstrates that the bionic TNS on the PS replica surface play a key role in its antireflective property. Thus, the bionic PS replica fabricated by μ-ICM not only inherits the nanopillars on the natural wing surface, but also its hydrophobic and antireflective properties.

The mechanisms for the aforementioned hydrophobicity and antireflectivity on the PS replica are analyzed as follows. Figure 12(a) schematically depicts the underlying mechanism for the surface wettability to a droplet. According to the Young's equation [32], the solid–liquid (γ_{SL}) and liquid–vapor (γ_{LV}) surface tensions attempt to make a droplet to shrink, whereas solid–vapour (γ_{SV}) tension attempts to make the droplet to spread out. For the smooth PS counterpart surface, $\gamma_{LV} + \gamma_{SV} > \gamma_{SL}$ due to the hydrophilic nature of PS, which means that it is energetically profitable for a droplet to spread on the surface rather than to shrink. So the PS counterpart surface exhibits a low CA (Figure 11(a)). As for the PS replica, the upper portion of the nanopillars is easy to be impregnated by water, and air is trapped in the gaps between the nanopillars to form a stable air pocket that acts as an effective water barrier. So, it is understandable that the PS replica exhibits an approximately superhydrophobic wetting sate (Figure 11(b)). Moreover, the solid–liquid–vapour three-phase contact line drops in the gaps between the nanopillars, and hence a global solid–vapour composite interface on the PS replica surface is formed. The droplets are pinned on the tilted PS replica surface with a moderate surface adhesion force (*F*). Therefore, a micro-droplet can be fixed on the PS replica surface because the *F* can overcome its gravity.

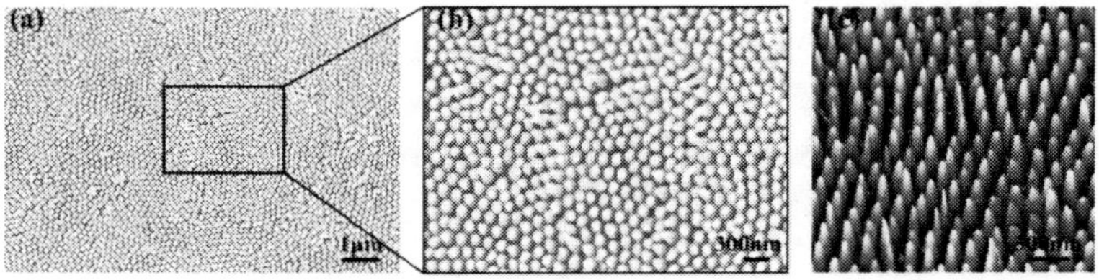

Figure 9. (a) Low and (b) high magnification SEM images and (c) AFM image of PS replica.

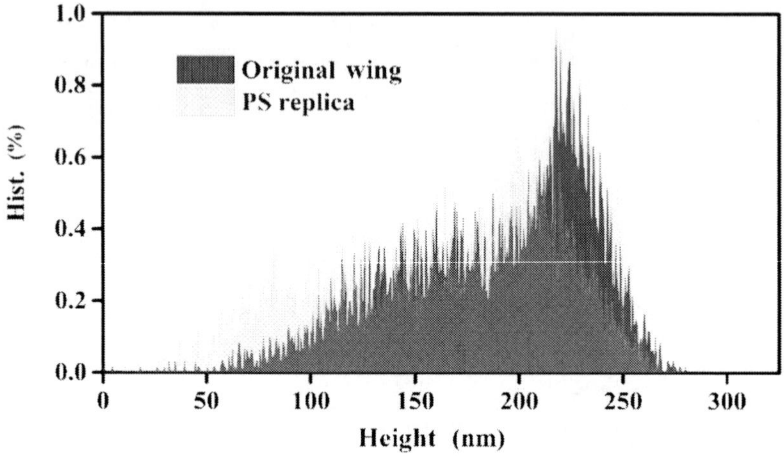

Figure 10. Height histograms of nanopillars on cicada wing and PS replica.

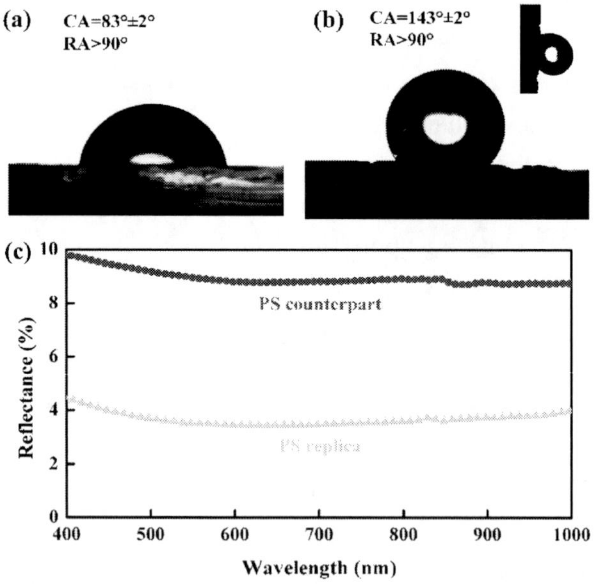

Figure 11. Wetting state of 4 μL droplet on (a) PS counterpart and (b) PS replica surfaces, and (c) reflectance spectra of PS counterpart and PS replica surfaces.

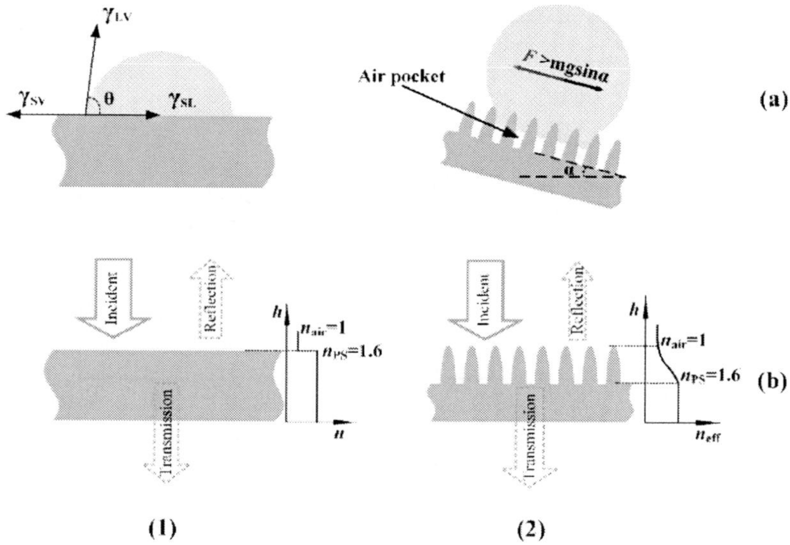

Figure 12. Schematics of mechanisms for wettability and antireflectivity of (1) PS counterpart and (2) PS replica.

Figure 12(b) schematically depicts the mechanism for the antireflectivities on the PS counterpart and replica surfaces. When a light wave is incident on the PS counterpart surface, Fresnel reflection occurs because of the difference in the refractive index (n) between the PS (n_{PS}) and air (n_{air}). The n changes abruptly at their interface, resulting in a high reflectivity on the PS counterpart surface. The nanopillars on the PS replica surface can be approximated as a set of multiple effective medium layers with a gradient refractive index ranging from the n_{air} to n_{PS}. According to the effective medium theory [33, 34], the effective refractive index (n_{eff}) for normal incidence passing through the nanopillars on the PS replica can be written as:

$$n_{eff}^2 = n_{air}^2 f + n_{PS}^2 (1 - f) \qquad (1)$$

$n_{eff}^2 = n_{air}^2 f + n_{PS}^2 (1 - f)$ (for reference only)

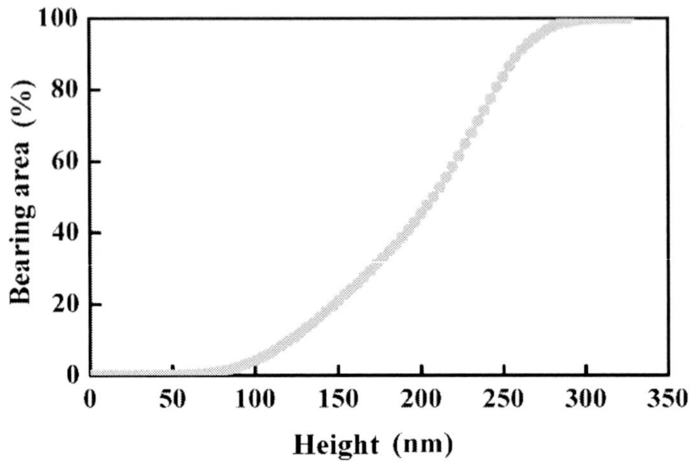

Figure 13. Bearing area distribution curve of nanopillars on PS replica surface.

where *f* is the fill factor of the nanopillars, which is the volume percentage of the nanopillars on the PS replica surface herein. According to the definition, the value of *f* is considered as the area ratio from bearing analysis [35] of the AFM image (Figure 9(c)). The area ratios along the nanopillar height were obtained from the AFM image by using the Nanoscope analysis software, and the result is shown in Figure 13. Accordingly, the n_{eff} along the nanopillars on the PS replica was calculated using Equation (1), and its distribution curve is plotted in Figure 12(b). It is obvious that the n_{eff} changes gradually from 1.0 (n_{air}) to 1.6 (n_{PS}) along the nanopillars. Therefore, the PS replica surface exhibits an obviously low reflectivity than the PS counterpart surface (Figure 11(c)).

CONCLUSION

The gradient wetting state along the veins on the natural cicada wing surface is revealed and the accumulated water droplets can be removed from the wing surface easily. This special property helps the cicada wing maintain a clean and light weight state. The nanopillars and gradient wetting states on the natural wing surface play a key role in the droplet transportation processes. This finding provides an inspiration for designing novel nanostructured surface with periodic or gradient wetting states, which may be developed into functional surfaces with antifogging, antibacterial, self-cleaning, and antireflective properties related to the natural properties of the cicada wing.

The nanopillars at the hydrophobic region on the natural cicada wing surface have a typical mean diameter of about 160 nm and a pitch of about 180 nm and play a key role in its hydrophobicity and antireflectivity. The orderly distributed tapered nanopores with a mean diameter of about 165 nm and a mean pitch of about 180 nm on the fabricated nickel replica are the inverse pattern of the nanopillars on the natural wing. Applying μ-ICM, the nanopillars on the natural wing surface are accurately and efficiently transferred to the PS replica surface with the nickel replica as the template. The shape, dimensions, and distribution of the nanopillars on the PS replica surface are consistent with those on the natural wing surface. The PS replica surface exhibits a CA of 143° ± 2°and an average reflectance of about 4% in the wavelength range of 400–1000 nm; thus, both hydrophobicity and the antireflectivity of the natural wing are well inherited by the bionic PS replica. Apparently, the composite water wetting interface and effective medium layer play a key role in the hydrophobicity and antireflectivity, respectively, of the bionic PS replica surface. The proposed strategy provides a desirable, straightforward, facile nanofabrication route for mimicking bio-inspired functional micro/nanostructures.

ACKNOWLEDGMENTS

Financial support provided by the National Natural Science Foundation of China (contract grant number: 51533003) and Guangdong Provincial Natural Science Foundation (contract grant number: 2016A030308018) is gratefully acknowledged.

References

[1] Bhushan, B. *Philos. Trans. R. Soc. A*, 2009, **367**, 1445–1486.

[2] Chen, H; Zhang, P; Zhang, L; Liu, H; Jiang, Y; Zhang, D; Han, Z; Jiang, L. *Nature*, 2016, **532**, 85–89.

[3] Han, W; Niu, S; Yang, M; Zhang, J; Yin, W; Ren, L. *Nanoscale*, 2013, **5**, 8500–8506.

[4] Gao, X; Yan, X; Yao, X; Xu, L; Zhang, K; Zhang, J; Yang, B; Jiang, L. *Adv. Mater.*, 2007, **19,** 2213–2217.

[5] Gao, X; Jiang, L. *Nature*, 2004, **432**, 36.

[6] Ko, DH; Tumbleston, JR; Henderson, KJ; Euliss, LE; DeSimone, JM; Lopez, R; Samulski, ET. *Soft Matter*, 2011, **7**, 6404–6407.

[7] Siddique, RH; Gomard, G; Hölscher, H. *Nat. Commun.*, 2015, **6**, 6909.

[8] Law, JB; Ng, AM; He, AY; Low, HY. *Langmuir*, 2014, **30**, 325–331.

[9] Koch, K; Barthlott, W. *Philos. Trans. R. Soc. A*, 2009, **367**, 1487–1509.

[10] Lee, LP; Szema, R. *Science*, 2005, **310**, 1148–1150.

[11] Dewan, R; Fischer, S; Meyer-Rochow, VB; Ozdemir, Y; Hamraz, S; Knipp, D. *Bioinspir Biomim.*, 2012, **7**, 016003.

[12] Kanamori, Y; Okochi, M; Hane, K. *Opt. Express*, 2013, **21**, 322–328.

[13] Meier, T; Solares, SD. *Nanoscale*, 2016, **8**, 17675–17685.

[14] Wang, CF; Hsueh, TW. *J. Phys. Chem. C*, 2014, **118**, 12399–12404.

[15] Jeong, KH; Kim, J; Lee, LP. *Science*, **2006**, 312, 557–561.

[16] Raut, HK; Dinachali, SS; Loke, YC; Ganesan, R; Ansahantwi, KK; Gora, A; Khoo, EH; Ganesh, VA; Saifullah, MSM; Ramakrishna, S. *ACS Nano*, 2015, **9**, 1305–1314.

[17] Xu, H; Lu, N; Shi, G; Qi, D; Yang, B; Li, H; Xu, W; Chi, L. *Langmuir*, 2011, **27**, 4963–4967.

[18] Kang, SH; Tai, TY; Fang, TH. *Curr. Appl. Phys.*, 2010, **10**, 625–630.

[19] Li, Y; Zhang, J; Yang, B. *Nano Today*, 2010, **5**, 117–127.

[20] Yu, X; Zhong, QZ; Yang, HC; Wan, LS; Xu, ZK. *J. Phys. Chem. C*, 2015, **119**, 3667–3673.

[21] Watson, GS; Myhra, S; Cribb, BW; Watson, JA. *Biophys. J.*, 2008, **94**, 3352–3360.

[22] Stoddart, PR; Cadusch, PJ; Boyce, TM; Erasmus, RM; Comins, JD. *Nanotechnology*, 2006, **17**, 680–686.

[23] Pogodin, S; Hasan, J; Baulin, VA; Webb, HK; Truong, VK; Phong, TH; Nguyen, V. Boshkovikj; Fluke, CJ; Watson, GS; Watson, JA; Crawford, RJ; Ivanova, EP. *Biophys. J.*, 2013, **104**, 835–840.

[24] Cai, J; Qi, L. *Mater. Horiz.*, 2015, **2**, 37–53.

[25] Sun, M; Watson, GS; Zheng, Y; Watson, JA; et al., *J. Exp. Biol.*, 212, (2009), 3148–3155.

[26] Zhang, GM; Zhang, J; Xie, GY; Liu, ZF; Shao, HB. *Small*, 2006, **2**, 1440–1443.

[27] Hong, SH; Hwang, J; Lee, H. *Nanotechnology*, 2009, **20**. 385303.

[28] Seol, ML; Woo, JH; Lee, DI; Im, H; Hur, J; Choi, YK. *Small*, 2014, **10**, 3887–3894.

[29] Chen, YC; Huang, ZS; Yang, H. *ACS Appl. Mater. Interfaces*, 2015, **7**, 25495–25505.

[30] Huang, YF; Jen, YJ; Chen, LC; Chen, KH; Chattopadhyay, S. *ACS Nano*, 2015, **9**, 301–311.

[31] Mo, X; Wu, Y; Zhang, J; Hang, T; Li, M. *Langmuir*, 2015, **31**, 10850–10858.

[32] Kwok, DY; Neumann, AW. *Adv. Colloid. Interface. Sci.*, 1999, **81**, 167–249.

[33] Ono, Y; Kimura, Y; Ohta, Y; Nishida, N. *Appl. Opt.*, 1987, **26**, 1142–1146.

[34] Chattopadhyay, S; Huang, YF; Jen, YJ; Ganguly, A; Chen, KH; Chen, LC. *Mater. Sci. Eng. R*, 2010, **69**, 1–35.

[35] Wu, F; Shi, G; Xu, H; Liu, L; Wang, Y; Qi, D; Lu, N. *ACS Appl. Mater. Interfaces*, 2013, **5**, 12799–12803.

INDEX

F

G

H

I

L

M

N

O

P

R

U

V

W

Z